"In *The Case for Space* Robert Zubrin lays out an exciting vision for the future of humanity in space. The narrative explains a practical, albeit challenging, path to the stars in the tradition of Mars Direct."

—John Grunsfeld, astrophysicist, astronaut,
Hubble repairman, and former NASA chief scientist
and associate administrator for science

"Zubrin is the leading space visionary of his generation, and the most eloquent."

—Stephen Baxter, author of *Voyage*, *The Time Ships*,
and *Vacuum Diagrams*

"In *The Case for Space* Zubrin shows why and how he is the foremost advocate for results, not mere rhetoric, in space. He applies a rigor to mission plans that NASA can only admire—though it doesn't—and try to emulate. His realism is the key to our larger future."

—Gregory Benford, author of *The Berlin Project*

"At last, here is the vision we were aiming for when we landed on the Moon. *The Case for Space* is the case for the future!"

—Buzz Aldrin, Apollo 11

"Zubrin is one of our generation's thought leaders in space exploration, and with *The Case for Space* he makes another major contribution to humanity's journey to the stars. We are several human generations since Apollo, yet nation-states have retreated from their once-bold path beyond our planet. With the amazing private efforts such as SpaceX now bearing fruit, a revolution is upon us such that space exploration is becoming the domain of private companies and citizens. Zubrin paints the picture of how the space frontier will be settled—in much the same way that the New World was—by individuals looking for new opportunities and new horizons. This is a bold and inspirational work drawn from Zubrin's insider knowledge of the industry and his own role in bringing this pioneering venture to fruition. This book is a must-read for every daydreamer, every space geek, and indeed every human being who dares to look up at the sky and imagine that someday we too will be ever present there."

—Jim Cantrell, chief executive officer of Vector Launch Inc.

"The tools needed to move into space have radically changed in the last decade. Zubrin has put together an excellent guide on how we got these amazing new machines and how we can now use them to break out into the solar system for the benefit of everyone. Highly recommended not only for space buffs but also for anyone who would enjoy an astonishing but true story about how a few somewhat eccentric individuals accomplished what neither NASA nor any other government space agency could do."

—Homer Hickam, author of *Rocket Boys*

"In *The Case for Space* Zubrin lays out a vision for our space program whose boldness and clarity of thought we have not seen since we took on the challenge of reaching for the Moon. This is the kind of thinking we need. Nothing great has ever been accomplished without courage."

—James R. Hansen, *New York Times*–bestselling author of
First Man: The Life of Neil A. Armstrong

"One thousand years from now, whatever the human race has evolved into, we will look back and see the next few decades as the moment in time that the human race irreversibly moved off the Earth and to the stars. Why should we do that? It is our moral obligation, and in *The Case for Space*, Zubrin creates one of the most powerful and coherent arguments to date. This is a must-read for anyone who cares about the survival of our species."

—Peter H. Diamandis, MD, founder and executive chairman of the
XPRIZE Foundation, executive founder of Singularity University,
and *New York Times*–bestselling author of *Abundance* and *Bold*

"Zubrin is a respected evangelist for manned spaceflight. This fascinating book—a distillation of his decades of expertise—combines a lucid exposition of future technologies with an eloquent and inspiring vision of humanity's future as a space-faring species."

—Martin Rees, Astronomer Royal, past president
of the Royal Society, and author of *On the Future*

"Beyond this tiny speck whirls a universe—truly everything. All who want their heirs to share some of that should read Zubrin's fact-filled call for a push beyond our planet. As it happens, out there lie the tools, resources, and keys to saving our earthly garden home."

—David Brin, bestselling, award-winning author of
The Postman, *Earth*, and *Existence*

THE CASE FOR SPACE

THE CASE FOR
SPACE

HOW THE REVOLUTION IN SPACEFLIGHT OPENS
UP A FUTURE OF LIMITLESS POSSIBILITY

ROBERT ZUBRIN

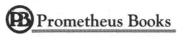

Prometheus Books

Guilford, Connecticut

Published 2019 by Prometheus Books

Cover image © Shutterstock
Cover design by Nicole Sommer-Lecht
Cover design © Prometheus Books

Inquiries should be addressed to
Prometheus Books
4501 Forbes Boulevard, Suite 200
Lanham, Maryland 20706
VOICE: 716–691–0133 • FAX: 716–691–0137
www.rowman.com

Distributed by NATIONAL BOOK NETWORK

Library of Congress Cataloging-in-Publication Data

Names: Zubrin, Robert, author.
Title: The case for space : how the revolution in spaceflight opens up a future of
    limitless possibility / by Robert Zubrin.
Description: Amherst, New York : Prometheus Books, 2019. | Includes index.
Identifiers: LCCN 2018061068 (print) | LCCN 2019000242 (ebook) |
        ISBN 9781633885356 (ebook) | ISBN 9781633885349 (hardcover)
Subjects: LCSH: Space colonies. | Outer space—Exploration. | Extraterrestrial bases. |
    Space industrialization. | Outer space—Exploration—Social aspects. Classification:
                LCC TL795.7 (ebook) | LCC TL795.7 .Z8285 2019 (print) |
        DDC 629.44/2—dc23
LC record available at https://lccn.loc.gov/2018061068

To Hope

CONTENTS

ACKNOWLEDGMENTS

I wish to acknowledge the contributions of my colleagues at Pioneer Astronautics—Mark Berggren, Stacy Carrera, Heather Rose, Jonathan Rasmussen, Max Shub, Jacob Romero, Steven Fatur, Andrew Miller, and Carie Fay—who are helping to develop many of the technologies discussed in this book. I also wish to acknowledge the help of my friends Richard Heidmann, Etienne Martinache, Richard Wagner, Freeman Dyson, and Mitchell Burnside Clapp, who reviewed early drafts of the manuscript or parts thereof and made many useful comments. Thanks are also due to Nicole Sommer-Lecht and Steven L. Mitchell of Prometheus Books for their excellent contributions with the book's art and editorial work, respectively. Special thanks are also due to my intrepid agent, Laurie Fox, who found an excellent publisher for the book. Most thanks of all must be given to my clever, funny, and darling wife, Hope Ann Zubrin, without whose patience, love, and constant support this book never could have been written.

INTRODUCTION

Great things are happening.

On February 6, 2018, the SpaceX Falcon Heavy rocket took flight, demonstrating a capacity to lift sixty tons to low Earth orbit and playfully sending a Tesla Roadster on a trajectory that took it beyond the orbit of Mars. To add to the coup, two of the Falcon's three booster stages flew back to land gracefully together at the Cape, while the third barely missed pulling off a recovery landing on a drone ship stationed downrange. (See plates 1, 2, and 3 in the photo insert.)

To understand how extraordinary this accomplishment was, let us recall that in 2009, the Obama administration's blue-ribbon review committee, headed by former Lockheed Martin CEO Norm Augustine, declared that NASA's moon program had to be canceled because the development of the necessary heavy-lift booster would take twelve years and $36 billion.

Yet SpaceX did it, in half the time and at a thirtieth of the cost. And, to cap it all, the launch vehicle is three-quarters reusable.

This is a revolution. The moon is now within reach. Mars is now within reach.

And it's just the beginning. SpaceX is developing the means to allow refueling the booster second stage after it reaches orbit. Once this technology is in hand, the Falcon's interplanetary payload will triple, giving it a capability greater than that of the mighty Saturn V rockets that sent astronauts to the moon in the 1960s. SpaceX's fully reusable 150-ton-to-orbit Starship launcher, now under development, will multiply that capability nearly three times over again. With such a system, the entire inner solar system will be wide-open to exploration and development.

The possibilities abound. A reusable rocket system that can send 60 or 150 tons to orbit can also send 60 or 150 tons from New York to Sydney in less than an hour. For comparison, a Boeing 737 has an empty weight of 45 tons. If the Falcon Heavy or its successor, the Starship two-stage launch system, can be made fully reusable, then an entirely new market for space launch can be created, one involving not a hundred or so launches per year, as is currently the case, but hundreds or even thousands *per day*. Such a market would drive a radical cheapening of space technology, finally making possible all the dreams of space tourism, industrialization, and colonization that have been within view but out of reach since the dawn of the space age.

So the dam has been broken, and the four-decade-long post-Apollo age of stagnation in space launch and human spaceflight technology has come to an end. An entrepreneurial space race has erupted with players including Firefly, Vector Launch, Virgin Galactic, Stratolaunch, and, most important, Jeff Bezos's Blue Origin—which will soon launch its own reusable New Glenn booster with similar capabilities to the Falcon Heavy—competing to take their share of a market that will soon explode in size. They will soon have plenty of company. SpaceX has shown that it is possible for lean, hard-driving entrepreneurial ventures to do—better—what it previously was thought only the governments of major powers could attempt. With the naysayers refuted, dozens of would-be emulators from around the world are sure to enter the fray, and their fierce competition will crash the cost of space launch and, with it, the cost of developing more advanced in-space technology as well.

Furthermore, while NASA and the other government space agencies may have gone adrift in the space launch and human spaceflight arenas, they have delivered extraordinary results in the fields of space science. Indeed, over the past several years, a series of remarkable discoveries has changed our understanding of the relationship between the human future and the rest of the universe. Launched in 2009, the wildly successful Kepler Space Telescope has found thousands of potential alternative Earths nearby. As a result, it is now clear that planetary systems—potential homes for life—are the rule in the universe rather than the exception. Also, since the 1990s, evidence has piled up to the point where it is now conclusive that asteroidal impacts on the Earth have been responsible not just for the extinction of the dinosaurs, but for other mass extinctions as well. The message here is that life on Earth is part of a larger cosmic system, which we humans ignore at our peril.

In 1994, the US Strategic Defense Initiative Organization (SDIO) launched a low-cost space probe to the Earth's moon, finding evidence for the presence of water—the staff of life and the basis of chemical industry—on the Earth's nearest neighbor. This was confirmed in 2009, when the Centaur upper stage that was used to launch NASA's LCROSS probe was crashed into a crater near the moon's south pole, sending up a cloud of water vapor. In 1996, NASA's Galileo probe uncovered evidence for what appears to be an *ocean* of liquid water under the ice-covered surface of Jupiter's moon Europa. This set off a wave of discoveries by subsequent probes, so that we now know that tidally heated under-ice oceans that may contain life are a general phe-nomenon, existing not only on several of Jupiter's other major moons but even Saturn's tiny satellite Enceladus—and probably on billions of other worlds throughout the universe.

A string of incredibly successful NASA orbiters and rovers launched since the mid-1990s has revealed Mars to be a world rich in all the resources needed to support life, and therefore future techno-logical civilizations. A few years ago, methane—which can only exist on Mars as a product of life or of hydrothermal environments that can support life—was detected by the Curiosity rover. Then, in 2018, sci-entists using the MARSIS ground-penetrating radar on the European Mars Express Orbiter announced the discovery of an underground lake of liquid salt water near the Martian south pole. On the basis of such data, it is becoming increasingly likely that we will discover not only the remains, but even living survivors of ancient microbial life on the Red planet.

As humble as such Martian microbes might be, the implications drawn from their existence are spectacular: the processes that lead to the origin of life are not peculiar to the Earth. If we combine this with the Kepler Mission's discovery that most stars have planets, and that virtually every star has a region surrounding it—near or far, depending upon the brightness of the star—that can support the type of liquid-water environments that gave birth to life on Earth and Mars, the conclusion is that a very large number of stars currently possess planets that have given rise to life.

The history of life on Earth is one of continual development from simple forms to more complex forms, with the more advanced forms manifesting ever-increasing degrees of activity, intelligence, and capability to evolve still further at an accelerated rate. If life is a general phenomenon in the cosmos, then so is intelligence. The implication is clear: we are not alone.

Yet there is still more good news. In early 2018, the SHARAD ground-penetrating radar team on NASA's Mars Reconnaissance Orbiter announced the discovery of massive ranges of glaciers on Mars, covered by only a few meters of dust, extending down from the poles to latitudes as far as 38 degrees north (the same latitude as San Francisco), and containing 150 trillion cubic meters of water, an amount six times greater than that contained in the Great Lakes on Earth. Not only that, the same team has also discovered gigantic underground caverns on the Red Planet—potentially offering vast volumes of shielded habitation for future human settlers.

Collectively, these discoveries are making it apparent what sort of birthright humanity has been given, if only we are bold enough to step up and embrace the challenge and the opportunity before us.

In *The Case for Space*, we explore the possibilities. Starting with a discussion of the present-day breakthroughs, we will take a deeper look at where it leads: to ultrafast global travel through suborbital space; to new industries on orbit; and to human settlement of the moon, Mars, the asteroids, the outer solar system, and ultimately the stars. All these things are possible, and we will explain how to achieve them.

Then we will look at what such mastery implies: what we will gain by undertaking this grand adventure and what we would lose by failing to do so. There is immense knowledge to be gained in space, but there are also deadly hazards to be faced, which we need to control if we are to protect ourselves and all other life on Earth. The value of the challenge itself, stimulating the creative forces of society as a whole and our youth in particular, will also be explored. There is also the question of the influence of an open frontier—or the lack thereof—on human freedom and on the battle of ideas that can sustain it or defeat it. Finally, there is the question of the human future itself: Will we be limited to one world, with limited resources and limited prospects? Or will we become something far grander in space, time, diversity, and ultimate potential?

Many years ago, the Russian space visionary Nikolai Kardashev outlined a schema for classifying civilizations.[1] According to Kardashev, a Type I civilization was one that had achieved full mastery of all the resources of its planet. A Type II civilization was one that had mastered its solar system, while a Type III civilization would be one that had control of the full potential of its galaxy. All of human history up to this point—from the trek out of our African birthplace to the settling of the continents and then the linking together of the disparate branches of humanity through first long-distance sailing ships, then telegraphs, telephones, radio, television, satellites, and the internet—has been a process of our rise from a local Kenyan biological curiosity to a full-fledged Type I civilization. That transition is now nearly complete, and we stand at the beginning of a new history—our rise to become a Type II civilization capable of measuring itself against the further challenge of becoming Type III.

It's a grand time to be alive. We are living at the beginning of history. We are present at the creation.

Nothing is inevitable, and nothing worthwhile is ever easy. Not all revolutions succeed. Some are suppressed by the forces of the old order. Others simply lose their way. We are surrounded by a living cosmos of unlimited possibilities. Will we ignore it or enter it? Will humanity retreat and allow itself to be, and to see itself, as mere passengers adrift in a sea of stars? Or will we step forward and, in taking hold of our solar system, take charge of our destiny, a species fully capable of contending with the challenges to come?

The choice is ours—yes, *ours*. We, the people of *this* time, this moment in history, have the privilege, the responsibility, and—provided we live up to the moment we have been given—potentially the honor and eternal glory of establishing humanity as a multiplanet, spacefaring species.

My father and all my uncles served in World War II. One of them landed on Normandy Beach. For doing such, they and their kind have rightly been called the Greatest Generation. By doing what history called upon them to do, they made this moment possible.

We owe it to them to make the most of it. We owe it to our posterity too.

It's our turn at bat. If we prove worthy, we can earn our own title as a Great Generation and be remembered as such for ages to come, on not just one but thousands of civilized worlds.

We can do it. This book will show you how.

Part I

HOW WE CAN

Chapter 1

BREAKING THE BONDS OF EARTH

Nothing can stop an idea whose time has come.
—Victor Hugo

December 21, 2015. The flight controllers are glued to their consoles. The SpaceX Falcon 9 took off minutes ago, and its upper stage is well on its way to deliver its payload of eleven ORBCOMM satellites to orbit. In the conventional sense, the mission is already in the groove to success. But the SpaceX team is reaching for more—much more. The Falcon's first stage is on its way down, and they are going to try to land it.

This isn't their first attempt. They tried five times before, and failed every time. No matter what they did, something always seemed to go wrong. The team is young—most members are under thirty, many just barely out of school. They had come to SpaceX and worked their hearts out—far, far harder than their peers who had found places at the established companies—in order to do what had never been done.

But maybe there was a reason why it had never been done.

Now, after ascending to more than one hundred miles altitude, the ship is coming back again for one more try. Hope mingles with dread as they once again hear the mission controller rattle off his narrative: "The first stage will soon begin its series of three burns to head back to Cape Canaveral."

But no burn is observed.

"This is bad. This is potentially bad," Elon Musk mutters.

Then the rocket lights and the team cheers, but only briefly, becoming somber again after the engine goes out. For three more minutes, they watch and wait. The second stage carrying the payload is driving hard toward orbit, but it is almost an afterthought. The real issue is what happens with stage 1, now coasting back toward the Cape with engines off. Will they light yet again?

"The second landing burn has commenced," the speaker announces.

The team cheers again, then quiets as they nervously watch the descent continue, the rocket's exhaust flame growing brighter and brighter as it screams down fast through the dark night sky.

A huge explosion sounds. Hearts sink: another failure. But no, telemetry keeps coming in. It isn't an explosion. It's the boom of the Falcon coming through the sound barrier.

The controller keeps talking.

"Final burn commencing."

"Altitude three hundred meters."

"Altitude one hundred meters."

Now the Falcon comes into view on the landing pad camera. It's descending slowly enough, but it's tilted! Will it tip over, and explode, as it had before? The rocket plume licks the pad. The team holds its breath, tense beyond measure. This is the moment.

"The Falcon has landed," the controller announces.

Musk yells, "It's standing up!" The room roars in wild cheers.

Staring at a computer screen video of the Falcon on the pad, Musk whispers, "Holy smokes, man." Then he sits down and taps out a tweet: "Welcome home, baby."

History turns a page.

Some people think that space settlement must be forever impossible because of its cost. This idea is absolute nonsense. There is no law of physics that requires spaceflight to be hopelessly expensive. On the contrary, the energy per unit mass that must be added to an object to send it from the ground to low Earth orbit is about nine kilowatt-hours per kilogram. At a typical American price of $0.08 per kilowatt-hour, this would amount to a cost of $72 to send an eighty-kilogram person with twenty kilograms of luggage to the space station. Of course, as in air travel, sending passengers also requires sending a flight vehicle that might weigh ten times as much. Even so, the energy to send a spacecraft to orbit is about the same as that required to fly a jet from Los Angeles to Sydney and back. Right now, round-trip tickets for such flights can readily be bought for under $2,000. Trips to low Earth orbit for the few private tourists who have been able to afford them have run ten thousand times that figure, with the bill to taxpayers for launching astronauts costing as much as ten times greater still. That's *$200 million* per passenger.

Figure 1.1. SpaceX Falcon 9 successfully lands at Cape Canaveral, December 21, 2015. *Image courtesy of SpaceX.*

Why is the price so high?

One obvious reason why space launch costs so much more than air travel is that aircraft are reusable, while space launch systems are expendable. A Boeing 747 seats four hundred people and costs $400 million. If the plane were expended after every flight, the ticket price for a round-trip would have to be at least $2 million each, or one thousand times the going rate. Of course, an expendable 747 could be manufactured more cheaply than one built to last, but even so, prices would still be wildly prohibitive.

Clearly, if space launch is to be made economical, the vehicles employed will have be reusable. Why aren't they?

There are two reasons for this: technical and institutional. I'll address the technical issues later. But suffice to say at this point that while significant, the engineering issues facing reusable spaceflight are all solvable. We've been flying to orbit for some sixty years now and have poured hundreds of billions of dollars into the field. We should have achieved fully reusable spaceflight decades ago.

Where there's a will, there's a way. Where there isn't a will, there isn't a way. It is not physics or engineering difficulties that bar our way to the cosmic frontier. To quote Shakespeare, the fault is not in our stars but in ourselves. It is we who, through our elected representatives, have created a set of institutional arrangements that have thus far blocked the achievement of that which we most desire.

The central institutional impediment to space progress is the system of cost-plus contracting the government has put into place in the very foolish belief that the price of hardware could best be kept under control by regulating contractors to charge their documented costs plus a modest set profit rate (of perhaps 8–10 percent). In fact, however, this system both forces contractors to increase their costs, by requiring legions of administrative personnel to document their billings, and incentivizes them to do so, since the more such overhead they incur, the more profit they make. This is in sharp contrast to the way business is generally done in the private sector, where customers are rightly interested not in how much it costs the manufacturer to produce an item, but only its quality and the price they need to pay to get it.

In the free enterprise world, manufacturers increase profit by cutting costs. In the cost-plus contractor world, manufacturers increase profit by increasing costs. No farmer or maker of widgets would ever staff up his or her operation with four administrative personnel for every field or factory worker. At major aerospace companies, this is done all the time. As a result, it is the norm for such contractors to have overhead rates exceeding 300 percent. Indeed, at the Martin Marietta company (later Lockheed Martin), where I was employed from the late 1980s through mid-1990s (and which was, along with Boeing, one of the two most successful of the eight major aerospace companies of that era), we at one point had more than thirteen thousand people at our primary facility, with fewer than one thousand working in the factory—leading one wit to scoff, "At Martin Marietta, overhead is our most important product."

Clearly, such a system is inimical to cost reduction. The government could probably cut its procurement costs in half nearly immediately, and perhaps as much as tenfold eventually, simply by getting rid of this insane system and buying things the way everyone else does. (This is not just a good idea. If the United States ever again gets into a fight with an enemy with comparable military capabilities, we are not going to be able to win with warplanes that cost as much as warships used to.)

It gets worse, as the inflation of the cost of space launch leads to the inflation of the cost of everything else associated with spaceflight. For example, if you are planning a communications satellite business and the cost of your launch to orbit is $100 million, it really doesn't matter much whether the cost of your satellite is $1 million or $20 million, because the difference only adds 20 percent to the capital cost of your operation overall. In fact, it probably drives your satellite cost over $50 million, because if you are paying $100 million just to get the satellite to its operating orbit, it damn well better work. So you are going to be more than willing to pay Cadillac prices for all your satellite's components and spare no expense in testing everything to the max. Moreover, if the satellite costs $50 million, the launch vehicle damn well better work too, so its cost will be sent up further as well, driving the cost spiral even higher. Furthermore, since neither the spacecraft nor the launch vehicle can afford to risk failure, only components that have previously been proven in spaceflight will generally be acceptable for use, a catch-22 that has led to colossal technological stagnation.

But the impact of cost-plus contracting goes far beyond inflating the cost of and impeding the improvement of spaceflight hardware items. It has fatally affected the progress of spaceflight overall.

For example, why should an industry that gets paid a multiple of its costs, whatever they may be, *want* to radically reduce the cost of space launch by introducing reusability?

The mainstream aerospace industry is not without merit. It rendered the nation magnificent service during World War II and, working at breakneck speed, got us to the moon. But that was in the days before cost-plus contracting and, moreover, was done by an industry management whose best patriotic impulses were being called forth into action by national leadership unmatched in quality, seriousness, and public spirit since that time. Even in the present period, the industry has displayed a professionalism and commitment to engineering excellence that has made NASA's robotic exploration and space astronomy programs—for example—epic successes. But reducing the cost of access to space is absolutely not one of its priorities.

But now, with the advent of entrepreneurial space launch companies led by visionaries who seek to reduce costs to the minimum—both because they are visionaries and because much of the money they are spending is their own—all that is about to change. Instead of running 300 percent overhead rates to inflate cost-plus billings to the government, these companies are ramming them down to the 20–30 percent typical of the commercial world. Instead of seeking business that allows them to bill for building a new rocket for every launch, these companies are developing reusable launch vehicles that will spread their construction costs across hundreds of launches. As a result, the hugely inflated space launch costs we see today are going to be radically deflated. The consequences of this will be world historic.

THE CONSEQUENCES OF CHEAP SPACE LAUNCH

How cheap can spaceflight get? As noted above, the energy to reach orbit is about the same as a round-trip flight from Los Angeles to Sydney: around $2,000 per passenger, or $20 per kilogram, including luggage. Air travel took a while before it could obtain such economies, so as a near-term goal for spaceflight, we'll estimate ten times that cost, or $200/kg to launch to low Earth orbit. To go to someplace interesting beyond low Earth orbit, we'll assume a cost ten times greater still, so $2,000/kg to go to the moon or Mars. As a check on these numbers, we note that it only takes about twenty-five kilograms of propellant to send one kilogram to orbit, and rocket propellant (for example, current kerosene/oxygen bipropellant—methane/oxygen would be cheaper) has a cost on the order of $0.40/kg. So the propellant cost of orbital launch is only about $10/kg—a figure that strongly supports the attainability of a $200/kg orbital delivery price for a reusable launch vehicle, with plenty of room for future improvement. What would this imply?

If the cost of space launch collapses, then the cost of all space hardware will drop, radically, because it will no longer be necessary to engineer all parts to superexacting conditions. Furthermore, the rate of progress of space technology will dramatically accelerate because it will no longer be necessary to confine designs to previously proven hardware. Not only that, there will be many more players doing things in space, including many risk-takers, inventors, and creative spirits of every kind, and so vastly more novel ideas will be tried and tested. The chains will be broken, the age of stagnation will end, and many of the long-deferred dreams of engineers will swiftly appear as reality.

But what will it mean for you, personally?

Well, for starters, it means that you would be able to fly to orbit for about $20,000. This is in the same range as current long-distance first-class airplane flights, taken by business executives and jet-setters insufficiently well-heeled to have their own private jets (which cost a lot more). Moreover, while such a ticket might seem a bit pricey for you and me, if we consider the probability that jobs on orbit are likely to be very high paying—easily north of $200,000 per year—the fare would be worth it, and most likely covered by employers as a small fraction of the compensation package. In short, with prices this low, not just a few ultrarich tourists but plenty of ordinary working stiffs would get to go to space, where, as we shall see in the next chapter, there will be plenty of work for them to do.

But space doesn't stop at low Earth orbit; that's just where it begins. Beyond Earth orbital space lies a vast frontier, and once prices are dropped to the level we are discussing, it will be wide open for human settlement.

Let's take Mars, for example. The Red Planet is a world with a surface area equal to all the continents of the Earth put together, containing all the resources necessary for life and technological civilization. But who will be able to afford to go there? At these transportation prices, anyone with a skill. One only has to look at American colonial history to understand this. Middle-class people, like many of the Pilgrims, paid for their one-way passage to America by liquidating their homes and farms. Common artisans, without such capital, paid their way by offering seven years' work in the new world. The modern-day equivalent, in either case, is about $300,000, roughly the same as a settler's ticket to Mars at the $2,000-per-kilogram rate.

In short, the day is not far off when the settlement of Mars will be as practical a proposition as the voyages of the *Mayflower* and the ships that followed her to create a New England in Massachusetts Bay.

Even Mars, however, is not the destination; it is simply the direction. Beyond Mars lies the asteroid belt, containing vast amounts of platinum-group metal resources with values of more than $40,000 per kilogram, making them exploitable at $200/kg launch costs. Their exploitation will make them much cheaper, to be sure, but will cheapen launch and deep space transportation costs as well—and, more to the point, make important new technologies such as fuel cells (whose expense is driven by platinum prices) economical. This will have enormous benefits for life on Earth.

Beyond the asteroids are the outer planets, whose atmospheres contain virtually unlimited resources of helium-3, the fuel for fusion reactors that can give humanity an endless supply of pollution-free energy and the means to take us to the stars.

All this can soon become attainable.

A new force has broken loose. A new tree is growing. We have only to water it, foster it, clear its way upward, and make sure that no one does anything to kill it.

Before the Renaissance, people believed the Earth was not only at the center—or rather bottom—of the universe but surrounded by crystal spheres made of unknowable, unbreakable material, forever barring us and banning us from the heavens.

The spheres are about to be broken.

GETTING TO $200 PER KILOGRAM

For the four decades following the Apollo program, the cost of launch to orbit held steady at about $10,000 per kilogram ($10 million per ton). With the advent of the reusable Falcon 9 and then Falcon Heavy, SpaceX has broken this barrier, crashing prices down to $2,000/kg. But can we reach $200/kg? In table 1.1, I give a general notion how this might be done.

TABLE 1.1. REDUCING LAUNCH COSTS

Booster	Payload	Stage 1	Stage 2	Other	Total	Price
Conventional	20 tons	$90M	$30M	$80M	$200M	$10,000/kg
Falcon 9 Expendable	20	60	20	40	$120M	$6,000/kg
Falcon 9 Reusable Stage 1	20	0	20	40	$60M	$3,000/kg
Falcon 9 Reusable	20	0	0	40	$40M	$2,000/kg
Starship	160	0	0	40	$40M	$250/kg
Falcon 9 Reusable 10X Flights	20	0	0	4	$4M	$200/kg
Starship 10X Flights	160	0	0	0	$4M	$25/kg

In the top line of table 1.1, we show a conventional aerospace industry cost-plus contractor booster, lifting twenty tons to orbit at a cost of about $200 million, or $10,000/kg. In the second line, I show estimates (I am not privy to SpaceX inside information) for the cost structure of the Falcon 9. Note that the first stage costs three times as much as the second stage. That is because it is nine times bigger, and a general rule in aerospace is that cost of hardware increases in proportion to the square root of its size. So the payoff of making the first stage reusable, shown on the third line, is quite large, cutting costs to less than a third of the conventional system. This is where the Falcon 9 is right now. SpaceX could advance this technology by making the Falcon 9 upper stage reusable, but since this stage is small, only a modest additional reduction in launch costs would be achieved by this advance because the other costs, such as payroll for the whole company and rents, dominate.

What to do? One approach is to go to a larger launch vehicle, such as the fully reusable heavy-lift Starship launcher that SpaceX has under development. By simply increasing the payload, a lower cost per kilogram can be achieved. Indeed, the Falcon Heavy, with sixty-ton-to-orbit capacity, already beats the Falcon 9, sporting a $2,000/kg price to orbit today.

But the largest potential gains lie in increasing the flight rate. With around thirty launches per year, a company like SpaceX spending something like $1.2 billion yearly would need to charge at least $40 million per launch to balance its books. But if it could get three hundred launches, only $4 million per launch would be needed.

Currently there are slightly fewer than one hundred satellite launches per year worldwide. So, for a single company to get three hundred launches, the total is going to have to be greatly increased. But this is exactly what lower launch costs could make possible.

THE POWER OF AN IDEA

There are much easier ways to make money than starting a revolutionary space launch company. Consequently, it will not be greed that propels us into space. It will be the power of ideas. Fortunately, the idea of creating a spacefaring future is a very forceful one.

Peter Diamandis is one of the most creative people working in the space community today. A former medical student turned serial entrepreneur, he has racked up a number of successes. One of the first (and least appreciated, because it brought him no money, but perhaps—as we shall see—the most consequential) was the founding of the intercollegiate organization Students for the Exploration and Development of Space (SEDS). Later, working with others, he created the International Space University, a notable institution with a substantial endowment that now has a campus in Strasbourg, France. In the 1990s, he decided to take on the central challenge of opening the space frontier—the creation of reusable space launch systems.

After reading a biography of Charles Lindbergh, Diamandis had been impressed by the fact that Lindbergh had flown the Atlantic—and eight others had been motivated to try—for the possibility of winning a prize. Furthermore, the estimated $400,000 (in 1920s money) collectively spent to win the Orteig Prize was much greater than the $25,000 prize fund itself. Not only that, the intellectual impact resulting from Lindbergh's spectacular achievement had resulted in the rapid expansion of the airline industry shortly thereafter.

Diamandis reasoned: if prizes had worked to unleash aviation, why not try the same approach for space? Inspired, he founded the XPRIZE Foundation, offering a $10 million prize to the first team that could send a reusable spacecraft to one hundred kilometers altitude (i.e., beyond the atmosphere, although not necessarily into orbit) twice within two weeks. This was a bit of a gamble for him, since he did not actually have the $10 million prize to give out. But he was willing to take that chance and ultimately found the funds when Iranian American billionaire Anousheh Ansari agreed to put up the cash. More than twenty teams entered the XPRIZE contest, and ultimately one of them, the Spaceship One group led by aviation genius Burt Rutan and backed financially by Microsoft cofounder Paul Allen, won the prize in 2004.[1] The headlines made by the success of Spaceship One (which was a small suborbital rocket plane launched off of a subsonic aircraft carrier vehicle) attracted Virgin Group CEO Sir Richard Branson, who adopted the concept and, with the addition of some truly serious money, has since been moving to commercialize it under the name Virgin Galactic as a system for suborbital space tourism. More recently, the late Paul Allen started his own company, called Stratolaunch, to commercialize a much larger version of the Spaceship One concept for orbital delivery.

There is a lot more that could be said about this episode, but the point

I want to make here is this: Ansari did not get, or expect to get, any significant financial return for her $10 million prize fund donation. Allen received the $10 million prize but spent $50 million to win it. Branson is spending hundreds of millions, and while it is possible he might someday see a profit from Virgin Galactic, he certainly knows plenty of better bets for making money. The same can be said for Allen and his Stratolaunch venture. In short, none of these people were moved to do these things by the power of money. They were moved by the power of an idea.

This brings me to Elon Musk.

In 1996, I published my book *The Case for Mars*, which laid out how we can reach Mars using present-day technology and why it is a societal imperative that we accept the challenge to do so.[2] This book was very successful, resulting in a deluge of more than four thousand letters and emails from a wide variety of people and, ultimately, the formation of an organization, known as the Mars Society, devoted to making the human exploration and settlement of the Red Planet a reality.

The Mars Society engages in public outreach, political work, and private projects, the most important of which have been the construction and operation of Mars analog research stations (for learning how to explore Mars on Earth) in both the Canadian Arctic and the American desert. This latter activity requires money. So it was that in the spring of 2001, we held a fundraiser in California's Silicon Valley.

The fundraiser entry fee was $500 per plate, but for some reason, one fellow sent in a check for $5,000. This got my attention. The check was signed by someone named Elon Musk. I had never heard of him. But I did some research and discovered that he was the founder of PayPal, a financial service that I *had* heard of, since a few irritating individuals kept asking us if they could pay their dues using it instead of credit cards or checks like normal people.

Under the circumstances, I thought it best to put this grievance aside and sought Musk out, meeting him for a very long cup of coffee before the fundraiser and then inviting him to spend a day with me at my company near Denver afterward. This kicked off a very productive interaction. Musk donated $100,000 to the Mars Society, which helped us fund the deployment of our Mars Desert Research Station, and he joined our board of directors for a while. He took a keen interest in the Mars Society's concept for the "Translife" (later "Mars Gravity") mission to fly a group of mice to orbit in a rotating capsule which would provide them with a long-duration life support system and a 38 percent gravity environment, thereby yielding the first data on how mammals—both those from Earth and those born on Mars— might thrive and develop in Martian gravity.[3] I hooked him up with a very sharp engineer I knew named Jim Cantrell to be his technical adviser, and together they went shopping for a low-cost launcher for the mission in Russia, where Cantrell had many contacts.

Some time later, however, Musk confided in me that he really didn't see himself as a team player helping someone else's operation. He had to have his own show—one lion on a hill and all that. Furthermore, he was wrestling with the question of what to do with the rest of his life. He had already made all the money he could ever want. Now he wanted to do something of enduring significance for humanity. He had read *The Case for Mars* and agreed with its thesis that transforming humanity into a spacefaring species was essential for the human future and the settlement of Mars was the key next step to bring that about. On the other hand, he also thought the creation of cheap solar energy was a critical advance that needed to be accomplished in our time. To which goal should he devote himself: Mars or solar energy?

I argued forcefully for Mars. Solar energy, I said, had obvious commercial potential. There were already billions of dollars flowing into it. Anyone with a credible idea for advancing it either technically or commercially could readily find investors. To that end, the technology would inevitably advance to its limit, and if that limit allowed solar energy to become more economical than fossil fuels, it would replace them, regardless of whether Musk was involved in the game or not. On the other hand, the business case for building a company to send humans to Mars is anything but clear. It would take a person of true vision to make it happen. If he did not devote himself to that goal, we might not make it.

In the end, Musk decided to do both, and he started an electric car company too. Thus was born SpaceX, the most remarkable aerospace company ever. When he began it in 2002, many space veterans shook their heads. *Another rich kid who thinks he can open the space frontier in his spare time; we've seen more than a few of them come and go*, most thought. Indeed, there had been some zillionaires behind several of the failed start-ups of the 1990s—Rotary Rocket and Beal Aerospace come to mind. But the backers of those companies were dilettantes. They threw a bit of spare change at some visionaries, and when things didn't work out or move fast enough, got bored and moved on.

Figure 1.2. The author and Elon Musk at the Mars Society convention in Pasadena in 2012. *Image courtesy of the Mars Society.*

But Musk was different. He didn't just put some money into SpaceX. He devoted his mind, his heart, and his full business talent. When I first met Musk in 2001, he had a good scientific background but didn't know anything about rocketry. When I visited him at his first factory in El Segundo in 2004, it was clear he had taught himself a great deal about rocket engineering—but he was still naive about the difficulty of space launch. When I told him he should expect his first several launches to fail, he brashly challenged my reasoning. But by 2007, he knew everything—including the pain of seeing his first two launches end in failure. He told me then he was good for one more try.

However, when the third launch failed too, Musk was tough enough to press on regardless, and in 2008, SpaceX finally reached orbit with the successful flight of its little Falcon 1—the first orbital launch vehicle ever developed by private money.[4]

In 2010, SpaceX bested this with orbital flights of Falcon 9, a bona fide medium-lift launch vehicle developed in one-third of the time at one-tenth of the cost generally considered normal by the mainstream, government-funded Fortune 500 launch companies.[5] But Musk didn't stop there. He developed the Dragon capsule, a vehicle capable of carrying human crews, with similar radically superior thrift and speed and then went on to demonstrate return, controlled landing, and reflight of the first stages on his rockets—a feat the mainstream aerospace industry has never been able to do, or try, at any price. Then, in 2018, he Sputniked the established aerospace industry again by flying the Falcon Heavy, a 75 percent reusable launch vehicle with twice the payload and one-third the launch cost of its closest competitor.

Musk spends SpaceX's money like it's his own—for the simple reason that much of it is. As a result, he does things much more cheaply than government-funded, cost-plus corporate contractors. He also does them more quickly, because in aerospace, cost equals people times time. He also possesses far more drive to innovate—to find new technologies that will make his spaceflight services ever cheaper still.

People sometimes ask me about Musk, since I knew him at the beginning of his quest. What is he like? He is a humanist, to be sure, but he is no Mother Teresa. There is a ruthlessness to his character. He doesn't hesitate to use people. There is also an aspect to his personality that some see as egomania. Indeed, if there is a practical flaw in Musk's personality, I would say that it is his difficulty in sharing credit with others. That is why the public has very little knowledge of the rest of the SpaceX team. This could cost him some of his best talent in the end. It also makes it difficult for him to join forces with others—for example, when, in 2013, billionaire Dennis Tito started his Inspiration Mars effort to launch a two-person Mars flyby mission, Musk gave him the cold shoulder.[6] But I don't think his aloofness is due to egomania. Rather, I think that what fundamentally drives Musk is a desire for what the ancient Greeks called *kleos*—eternal glory for doing great deeds. He enjoys money and finds it useful, but he's most definitely not in it for the money. He doesn't want fame as such—cheap fame of the Paris Hilton variety has no interest for him. No, he wants kleos.

In Shakespeare's *Henry V*, on the dawn of the day of battle, the king says:

The fewer men, the greater share of honour.
God's will, I pray thee wish not one man more.
By Jove, I am not covetous for gold . . .
But if be a sin to covet honour
I am the most offending soul alive.

That's Musk to a T. Not an ideal person, perhaps, but of such stuff are heroes made.

Musk wants to open the space frontier. But if the only competition he faced were that offered by the wildly expensive cost-plus aerospace majors, the temptation to just take their business and make off like a bandit by simply pricing his launches at a modest discount against theirs—rather than cheaper by a factor of ten or more—would be very strong indeed. This would be particularly true if Musk were to go public and had to explain his business decisions to dividend-driven investors. Fortunately, he's not going to have that option for long.

Where Musk has gone, others are already following.

The most important of the SpaceX emulators is the Blue Origin company, founded by Amazon CEO Jeff Bezos. Once again, we see the power of the space idea. I had no direct involvement with the chain of events that led to this rather secretive start-up, so I don't know all the details, but it appears that the key formative influence was Princeton professor Gerard K. O'Neill, whose visionary concepts, published in the 1970s, of building orbiting space cities financed by solar power satellites beaming power down to Earth inspired a large following.[7] Among these followers was the young Bezos, who included many O'Neillian ideas in his high school graduation address and then went to Princeton, where he studied with O'Neill directly. While at Princeton, Bezos was also recruited into Peter Diamandis's SEDS organization, which further expanded and deepened his contacts in the space movement, who no doubt contributed further to bringing him into the fold. In any case, hats off to all who helped deliver the message, because Bezos is reportedly the richest man in the world, and as forces on behalf of the spacefaring future go, they don't come much bigger.

Blue Origin's goal is also reusable space launch, and it has demonstrated significant steps toward achieving that goal by repeatedly performing controlled landing of the booster stage of its suborbital (reaching one hundred kilometers in altitude) New Shepard launch vehicle. Until late 2016, Blue Origin's public profile was that of a company planning to offer suborbital five-minute zero-gravity rides to space tourists, which would not have made it a serious competitor to SpaceX or the aerospace majors. But in September 2016, Blue Origin announced its plan to create the New Glenn, a reusable two-stage Earth-to-orbit booster. Bezos says he is prepared to put $1 billion per year of his own money into Blue Origin to make this system a reality, so it must be taken very seriously.[8]

SpaceX's Falcon Heavy employs a triplet of the same kerosene/ oxygen first stages that power its veteran Falcon 9 medium lifter. In contrast, New Glenn is a totally new system that will use a single large-core methane/oxygen booster for takeoff. But while the designs of the two systems are completely different, their fifty- to sixty-ton-to-orbit delivery capacities (about twice that of the much more expensive Atlas and Delta systems offered by the Lockheed Martin–Boeing United Launch Alliance) are close to identical, making them direct competitors. So the new space race is on.

They won't be its only participants. Around the world, people are watching these developments and planning their own entries. Perhaps nowhere is the desire to emulate Musk's accomplishments felt more fiercely than in Russia, a country with a very proud spaceflight tradition of its own. The old Soviet space industry still exists, for the most part, and by sticking to designs that have worked since the 1960s, it can punch out launches at somewhat lower costs—if significantly lower reliability—than the American aerospace majors. But Russia has nothing like SpaceX—yet. I am aware of groups over there who would like to start such an enterprise. It's possible they could pull it off. The country certainly has the aerospace talent, and there are large pools of capital—some of which are held by people who share the vision of human expansion into space—potentially available to support such a venture. The only thing standing in their way is the kleptocratic governing style of the rulers of the Kremlin, which introduces uncertainties into any investment: why make what others might just take? But, as we shall discuss in the next chapter, Putin and company must know that if they are to maintain anything like military parity with the West, they are going to need comparable space launch capabilities. Like it or not, they may have to reform.[9]

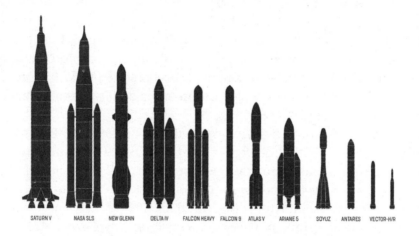

SATURN V   NASA SLS   NEW GLENN   DELTA IV   FALCON HEAVY   FALCON 9   ATLAS V   ARIANE 5   SOYUZ   ANTARES   VECTOR-H/R

Figure 1.3. Space launch systems. The key reusable commercial competitors for the near term are the SpaceX Falcons and the Blue Origin New Glenn series. The United Launch Alliance's veteran Atlas and Delta series are being priced out of the market, a trend ULA hopes to correct with its new Vulcan booster. The Russians and French are going to have to come up with something better than their expendable Soyuz, Proton, and Ariane rockets soon if they want to remain competitive. The NASA SLS offers almost twice the lift capability of the Falcon Heavy or New Glenn, but its costs are an order of magnitude greater. The Saturn V moon rocket could lift even more (140 tons to orbit) but went out of production almost half a century ago. At the small end, there is a new generation of entrepreneurial microsatellite launchers, typified by the Vector-R and the Electron-sized Vector-H. *Image courtesy of Kim Jennett, Vector Launch.*

Modernity doesn't just offer a carrot; it also holds a stick. The need to compete with other nations forced literacy on Russia. Who knows, it might yet force liberty as well.

But an entrepreneurial space race need not be limited to entrants from countries that already boast major government space establishments. People can enter from anywhere—and they will—provided that their countries offer decent conditions, educations, skills, and property protections and a fair amount of liberty under law.

To paraphrase John F. Kennedy's words from the dawn of the space age: a new ocean has opened, and free people will sail it.

FOCUS SECTION: RISE OF THE MICROLAUNCHERS

The entrepreneurial space revolution has not just opened up the new frontier to billionaires. It has made it possible for inventive people of ordinary means from all over the world to obtain investment to start their own space launch companies. Dozens of such ventures have been launched, with many securing substantial funding. Three of the most promising are Vector Launch, Firefly, and Rocket Lab.

Vector Launch was founded by Jim Cantrell, the crack engineer I hooked up with Elon Musk to help him launch SpaceX. After getting that venture underway, Cantrell and his friend Jim Garvey split off to start their own outfit, focusing on the microsatellite market, which is becoming increasingly important as advances in electronics make possible tiny spacecraft with capabilities equaling those of multiton satellites of a generation ago. Accordingly, the company's first vehicle, the Vector-R, is being designed to deliver a payload of just sixty kilograms to orbit (compared to twenty-two thousand kilograms for a Falcon 9). Among Vector-R's innovative features is the designers' decision to use LOX/propylene propellant. This is a combination that my team at Pioneer Astronautics, backed by NASA funding, first put to the test using small-scale rocket engines in 2003. We found it to be highly attractive, as it delivered a significantly higher exhaust velocity than the time-honored LOX/kerosene rockets used by SpaceX and Lockheed Martin (3.7 kilometers per second, compared to 3.4), and was easier to start and restart as well. Our test engines had a thrust of one hundred pounds. Vector Launch has developed practical units delivering six thousand pounds of thrust; designed a complete two-stage launch vehicle to employ these; built a launch site in Kodiak, Alaska; and made progress on numerous other fronts. As a result, in late 2018, the company received $70 million in investment. First launch is expected in 2020.

Firefly was founded by Tom Markusic, a veteran of NASA, Virgin Galactic, Blue Origin, and SpaceX. Designed to deliver one thousand kilograms to orbit, Firefly Alpha is a two-stage booster combining traditional LOX/kerosene propulsion with innovative carbon structure technology. In 2016, Firefly came close to bankruptcy but was rescued by Ukrainian American investor Max Polyakov, who then split its operations between the United States and Dnipro, Ukraine, where a great deal of aerospace talent is available at bargain prices. As a result, Ukraine may well beat Russia into the entrepreneurial space race.

New Zealand–based Rocket Lab has done best of all the micro-launcher start-ups. Employing revolutionary carbon structural technology; advanced 3-D printed, electrically pumped rocket LOX/kerosene engines; and a host of other inventions, the company's Electron launch vehicle reached space on its first test flight in May 2017 and succeeded in delivering satellites to orbit in its second test flight in January 2018.

As a result, orders are now pouring in.

(See plate 4.)

Chapter 2

FREE SPACE

Reusable rockets fielded by competitive entrepreneurial companies are necessary to bring the cost of space launch down to levels that can truly open the cosmos to humanity. But they are not enough. One factor is missing: the launch rate must be radically increased.

To understand why this is so, let us consider the situation facing SpaceX, the leanest, meanest, smartest, and by far most cost-effective space launch company around. SpaceX currently has around six thousand employees. At an estimated average yearly cost per employee, including wages, benefits, and taxes of $100,000 each, that comes to a payroll bill of $600 million per year. If we assume all other company costs, including materials, rents, taxes, insurance, legal fees, and others, are at least that much again, the total gross annual bill for running the company must exceed $1.2 billion.

Now, during 2018, SpaceX performed twenty-one launches. This was an incredible feat because not only were they successful every time, but the number of launches constituted 20 percent of the entire world launch market—truly an extraordinary result for a single, relatively small company. But even so, if we take the $1.2 billion low estimate bill to run the company, divide it by twenty-one launches, and assume (rather naively) that all Elon Musk wants to do is break even, we find that SpaceX would need to charge at least $60 million per launch to stay afloat. In fact, they charge about $80 million—or about $4,000 per kilogram for a twenty-ton payload. This is less than half the price anyone else is offering, but it's no revolution in launch costs.

It's a matter of simple math. Even with full reusability, if we want to get the cost per launch down by a factor of a hundred, we are going to need a market for at least a hundred times as many launches. Can such a potential market be created?

I believe it can. It won't be led by satellite launch—that market (about one hundred launches per year globally at last count) is far too small, even if we take into account its potential tripling as a result of the massive satellite constellations that Musk and others are now planning. But there is a much larger space launch market waiting to be opened, and that is long-distance rapid passenger travel around the Earth.

A fully reusable orbital class launch system could be used to deliver passengers point to point from anywhere on the surface of the Earth to anywhere else in less than an hour. As someone who has done my share of twelve-hour-plus global air flights, I can testify that the ability to cut short their monotony could be worth a lot. With my middle-class means, I just suck it up and fly economy. But others pay up to $20,000 per ticket to make such trips less unpleasant.

So consider one such route: Los Angeles to Sydney. Currently it is serviced by many flights per day of large jets, each taking eighteen hours. If a rocket plane service provided just one flight per day each way, that *single route alone* would increase the world launch market by 730 flights, multiplying it eightfold. But what about Seattle to Sydney, New York to Sydney, Atlanta to Sydney, London to Sydney, London to Johannesburg, London to Rio, New York to Rio, New York to Abu Dhabi, St. Petersburg to Rio, Tokyo to Santiago, New York to Tokyo, Atlanta to Shanghai, Los Angeles to Bombay, and so on, and so on? There are dozens of worthwhile routes, collectively providing a market for tens of thousands of flights per year.

But is the technology near at hand to make such transportation systems possible? It wasn't before, but it is now. The key to global transport using rocket propulsion is a two-stage, fully reusable system in which the first stage returns to the launch site and the second makes the long-distance trip, to be sent home with the help of a first stage based at the destination. This is precisely the sort of system that is now emerging from the two-stage reusable booster development programs of SpaceX and Blue Origin.

The reason a two-stage system is necessary is a result of the basic equations of rocketry. To obtain global reach, a rocket must reach orbital velocity, which in the case of the Earth is eight kilometers per second. However, a rocket experiences velocity loss during ascent due to aerodynamic drag and gravity, and so the real velocity increment (or "delta-V," abbreviated ΔV) that needs to be delivered by its propulsion system is closer to 9.5 km/s. This is more than twice the exhaust velocity of any practical propellant combination, which means that the amount of propellant needed will greatly outweigh the payload. (See box.)

As the amount of propellant increases, however, the vehicle needs ever-bigger tanks and engines, which eat up ever-larger fractions of the dry mass until there is nothing left for payload. In consequence, it is impossible for a single-stage rocket system to obtain the 9.5 km/s ΔV orbital velocity necessary for global reach, because its payload falls to less than zero. It could go some distance with a small but nevertheless positive payload, but when all is said and done, the potential performance is insufficient. This is shown in figure 2.1, which compares the payload and range of a single-stage rocket with that of a two-stage system of the SpaceX/Blue Origin type, where the stay-near-home first stage does the first 4 km/s of ΔV, while the passenger-carrying upper-stage vehicle does the rest. In figure 2.1, the rocket planes are all assumed to have a ground liftoff mass of 2,500 tons, roughly the takeoff mass of the Saturn V, space shuttle, or SpaceX's Starship booster system design, and employ methane/oxygen rockets for propulsion.[1] (Methane/oxygen is the best propellant combination for rocket planes because it offers high performance, ease of handling, worldwide availability, and very low cost.) Their payload, in tons, is shown on the vertical axis as a function of range, in kilometers, which is displayed on the horizontal axis.

It can be seen that the payload of a single-stage rocket plane falls to zero at a range of eight thousand kilometer, or five thousand miles, while the two-stage system can deliver a one-hundred-ton payload passenger cabin over global distances. Figure 2.1 shows performance data for both ballistic flight vehicles and winged craft with lift/drag ratios of 2 (the space shuttle had a hypersonic L/D of 1, NASA's Orbital Sciences X-34 of 2.5). Such lifting configurations allow gliding after reentry, but they add dry mass to the system, so the net range extension they provide is limited. Higher L/D winged configurations are possible that could enable somewhat greater reach for each option, but the basic story remains the same.

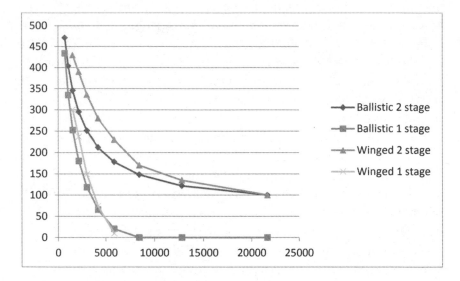

Figure 2.1. Comparison of the payload of two-stage and one-stage reusable rocket planes as a function of range. Payload (*y* axis) is in tons, range (*x* axis) is in kilometers.

So what would this two-stage global reach rocket plane be like? Takeoff would certainly be noisy, as would landing for nonwinged options, so the pads for such systems would probably have to be platforms located some tens of kilometers offshore, or else far out in open country or desert. This suggests a brief boat, seaplane, or helicopter ride would also be part of the trip. Also, an aspect of the experience of a transglobal flight would be about forty minutes of zero gravity and the same view of space that astronauts get. This would be a big plus for sales. In fact, companies like Virgin Galactic are currently offering four minutes of zero-gravity experience for $200,000, without transporting anyone anywhere, and getting a fair number of takers.[2]

By my math, each flight of a one-hundred-ton passenger cabin rocket plane would consume about 2,100 tons of methane/oxygen propellant. At a cost of $120 per ton, this would entail a fuel bill of about $250,000 per trip. At 140 tons, including cabin, tanks, and engines, the long-distance flight vehicle would have a dry mass comparable to a Boeing 767, which carries about two hundred passengers. If each passenger paid the $20,000 price of current global-distance first-class tickets (getting, in addition to fast global transportation, ten times the Virgin Galactic zero-gravity fun for one-tenth the price), a gross revenue of $4 million per flight could be obtained. That leaves plenty of room for other operating costs besides propellant, even allowing for some tickets to go for less than the premium price. (At $5,000, I'm in, for at least one flight, because I want to experience zero gravity and the brilliant starry sky of space at least once before I die. Book me for New Zealand.)

TECHNICAL NOTE: FUNDAMENTALS OF ROCKETRY

Let's say you weigh fifty kilograms and are standing on roller skates. If you throw a five-kilogram brick in one direction with a velocity of 10 meters per second, that action will send you scooting the other way with a velocity of 1 m/s. This illustrates the basic principle known to physics as conservation of momentum. Split an object into two parts and send them flying in opposite directions: the momentum—or mass times velocity—of each will be the same.

Rockets work on this same principle. The more momentum a rocket vehicle can shoot out one way in the form of fast-moving propellant gases, the more it can increase its own velocity the other way. In the example above, the skater weighs more than the brick and so moves away slower. A rocket might work that way too, using a small amount of propellant to effect a tiny velocity change. If speed is called for, however, a rocket can be made to move faster than its propellant exhaust velocity by piling on the gas. But while the final kilogram of propellant used by a rocket only has to push the rocket, the next to last has to push both the rocket and the final kilogram, and the third to last must push all the above, and so forth. As a result, the amount of propellant a rocket needs to use increases exponentially with the ultimate speed desired. This leads to the famous rocket equation, which says that for any rocket with exhaust velocity C, the "mass ratio," the ratio of its total mass (i.e., its dry mass plus propellant) to its dry mass alone increases exponentially with the factor $(\Delta V/C)$. Put mathematically:

$$\text{Mass Ratio = Total Mass/Dry Mass} = e^{\Delta V/C} \qquad (1)$$

The meaning of this equation is shown in figure 2.2, below, where we see how the mass ratio of the vehicle and the payload it can deliver changes as the key factor $\Delta V/C$, displayed on the horizontal axis, changes.

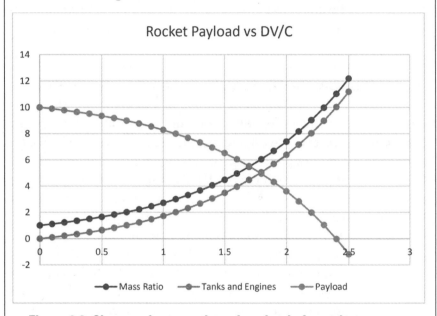

Figure 2.2. Change of mass ratio and payload of a rocket as a function of $\Delta V/C$.

In figure 2.2, it is assumed that the vehicle has a dry mass of ten tons, including payload, tanks, and engines. If $\Delta V/C$ is zero, no tanks or engines are needed, so the ten tons of dry mass is all payload. But as $\Delta V/C$ increases, the mass ratio goes up, and more and more propellant is required. So now the vehicle needs tanks and engines, and these might typically weigh 10 percent of the propellant, which is what I assume in figure 2.2. But the vehicle only has ten tons of dry mass, so as the tanks and engines get bigger, the payload must get smaller, falling drastically when $\Delta V/C$ is greater than 2.0, until mission capability drops to zero when $\Delta V/C$ reaches 2.4.

Methane/oxygen rocket propellant has an exhaust velocity of

3.7 km/s, so for an orbital mission where ΔV is 9.5 km/s, $\Delta V/C$ is 2.6, making a single-stage-to-orbit vehicle infeasible. Hydrogen/oxygen can obtain an exhaust velocity of 4.4 km/s, so $\Delta V/C$ using it would only equal 2.16. But that propellant is not only much more expensive but bulkier, which increases the mass of the tanks, so at the end of the day, a vehicle employing H_2/O_2 would do no better.

(Note: Astronautical engineers frequently report the exhaust velocity of rockets not in units of km/s but in seconds of "specific impulse," or "Isp." Conceptually, the specific impulse of a rocket is the number of seconds it can deliver a pound of thrust using a pound of propellant. That said, you can translate Isp, given in seconds, directly to exhaust velocity, given in meters per second, simply by multiplying the Isp by 9.8. So, for example, a typical methane/oxygen rocket engine with an Isp of 378 seconds would have an exhaust velocity of 3,704 m/s or 3.7 km/s.)

If we use two stages, however, the required ΔV can be split between them, with the $\Delta V/C$ required of each being less than an easy-to-do 1.4.

This is why the two-stage reusable vehicles being developed by SpaceX and Blue Origin are the right designs for both orbital delivery and fast intercontinental travel.

More broadly, if you want any rocket vehicle to achieve a ΔV more than twice its exhaust velocity, you need to use more than one stage.

SPACE TOURISM

Since the 1990s, a number of promoters have pointed to space tourism as a clever business path to opening the final frontier. In the era of ultra-high-priced spaceflight, such concepts were impractical, with the only achievements being several flights delivering a few billionaires to either the Russian Mir or the International Space Station at prices of around $20 million each.[3] However, in the relatively near future this picture could change radically, as the same intercontinental rocket plane technology enabling fast global travel could also be used to send passengers to orbit at much more affordable rates.

Why would anyone want to take a vacation on orbit? Well, for those who have had too much of the Aegean islands, Aspen, and Tahiti, a stay in a space hotel could offer something truly different. Still not convinced? How about the attractions of zero gravity, which, it has been argued, will be of special interest to honeymooners and other fun-loving couples? (There is a hilarious folk song popular in the science fiction community in which exactly the opposite proves to be the case.) This experience could well be expected to be enhanced, at least for some people, if the bedroom suite module includes a huge transparent window facing downward to give the couple a spectacular view of the blue rotating Earth (and vice versa). For those with other tastes, the module window could face out toward the endless sea of space with its myriad of unblurred stars glistening like a million jewels on black velvet. In between bouts in the bedroom, the couple could enjoy unique zero-gravity sports such as tennis, racquetball, basketball, soccer, gymnastics, or martial arts carried out in a large module suitably designed to accommodate numbers of people rapidly bouncing off the walls. For a modest extra charge, guests could take classes in extravehicular activities and become certified to wear space suits and go EVA. An astronaut certification suitable for framing would also be provided. Those with a more sedentary bent could while away the hours before bedtime engaged in astronomy or Earth studies in the hotel's observatory. To increase the variety offered by the hotel's primary attraction, a matchmaking service could also be provided. This would be especially valuable since in addition to being fun-loving and adventurous, most of the people you would meet at the hotel would undoubtedly be rich.

It's easy to see how such a business could evolve once there are orbital rocket planes engaged in fast global transportation. For example, instead of flying from New York to Sydney and landing, the rocket plane could remain in space for several orbits, turning a one-hour trip into an all-day trip, before landing back at the Big Apple. A logical evolution of such excursions might be to extend the day trips to overnights and eventually to weeklong cruises. But the cruise ship model only goes so far, because a rocket plane that could do a surface-to-surface flight every day with two hundred passengers would have to increase its rates quite a bit to maintain income when switching to weekly cruises, each necessarily accommodating fewer people.

A better plan, therefore, if multiday trips are contemplated, would be to create space hotels that remain on orbit with all the accommodations appropriate for a fun vacation and only use the rocket planes to ferry customers up and take them home.

Such hotels are already in the works. Robert Bigelow, the billionaire owner of Budget Hotels of America, has founded a company called Bigelow Aerospace to build them and has already tested habitation module technology in space.[4] When Bigelow founded his company some twenty years ago, he was way ahead of his time. But now it would appear that time may soon catch up with him.

DOING RESEARCH ON ORBIT

Orbital research labs that take advantage of the unique zero-gravity and high-vacuum environments available in low Earth orbit could also be producing a profit in the relatively near future. The product of such labs is knowledge, which is massless. Thus, precious little raw material is required, at least in principle, to produce marketable products of enormous cash value. This is so forcefully the case that even the vastly overpriced space shuttle managed to produce what might be considered a kind of profit during two ten-day missions in which zero-gravity experiments helped researchers determine the structure of certain animal viruses, thereby enabling the development of veterinary vaccines worth several billions to the economy. As part of the shuttle program, a successful company called SpaceHab created and rented out the use of its lab module, which flew periodically on the shuttle. Because of the high costs and difficulty of dealing with the opaque NASA bureaucracy, commercial enterprise never signed up for this service in a big way, leaving NASA as the primary customer. The same proved true of the research facilities offered by the International Space Station. However, a dedicated orbiting research lab with a long-duration professional staff could offer much more than simply lab space. With a lower-cost launch vehicle to support the operation, prices could conceivably drop to the point where investment in such research would be competitive with the return offered by terrestrial research facilities.

On orbit, the distorting influence of gravity is nearly absent, which creates conditions enabling the production and determination of the structure of various types of crystals and other compounds. In addition, low Earth orbit provides access to very high-quality vacuum conditions that cannot be economically produced in earthbound labs. The knowledge that flows from investigations conducted in these environments can enable the development of a range of products, from disease cures to "brains" for new supercomputers so advanced that their proponents claim they would revolutionize life on Earth. Potential microgravity or high-vacuum research products with astronomical value exist in the form of vaccines, synthetic collagen (which could be used to construct corneas), targetable pharmaceuticals, structured proteins, crystal materials (for computer chips and quantum devices), ultrapure epitaxial film production, unique polymers and alloys, and electrophoresis applications. These products could lead to breakthrough applications in such high-growth areas as semiconductors, computers, instruments, biotechnology, and drug manufacturing, areas that today represent a business base of more than $240 billion per year.

Some have advanced the notion that zero-gravity research could better be done on unmanned satellites. No doubt such facilities will also be launched. But having run a research lab myself, I believe that such claims are mistaken. Yes, isolated well-planned experiments can be flown on automated spacecraft and useful data returned. But to effectively perform an investigative research program into unknown intellectual territory requires real live human experimenters with constant access to their apparatus. An automated experiment can record data that are expected; only a human investigator can respond to surprises—and most big discoveries come as surprises.

The right surprise could be worth billions.

ORBITAL INDUSTRIES

Producing patents is the best way for an orbiting lab to make money. The discovery of knowledge in space that enables industrial processes to be realized on Earth is clearly the highest payoff path for such space-based facilities. But what if that is not possible? What if the lab discovers a process that can only be replicated in the zero-gravity environment of space? Could profitable mass-production operations actually be initiated on orbit?

The answer to this question depends upon a variety of factors, chief among them the cost of space launch, which today stands at roughly $5,000 per kilogram to low Earth orbit. At such rates, creating orbital industries is out of the question. But at the $200/kg rate that now appears to be achievable, matters change radically. In order for space-based manufacturing to be profitable, the value of the goods produced per unit weight must exceed this figure. In fact, it must exceed it by a good deal, because in addition to transporting the raw materials up and the product back down, the launch system will also have to transport the orbital factory; its spares, consumables, and power system; the workforce and their consumables; and the propellants and other consumables necessary to keep the factory spacecraft functioning and stable in its proper orbit. In addition, the orbital factory business will have to support the salaries and fringe costs of the company's earthbound and spacebound staffs; its offices, advertising, insurance, taxes, interest payments, and other overhead; and the standard retail markup, and, given the high level of risk in such a business, will have to pay large dividends to investors. So if the launch cost is $200/kg, the orbital factory's product will have to boast a *retail* sales price of at least $2,000/kg for there to be a net payoff sufficient to motivate investment. Roughly speaking, this is about one-twentieth the price of gold or three times the price of silver ($2,000/kg works out to $57/ounce). In addition, the product produced would have to be so superior to terrestrial alternatives that it would sell well despite the fact that it might cost more. Taken together, these factors would tend to rule out almost all alloy or other materials production operations, but the production of advanced computer chips, unique pharmaceuticals, or lossless fiber-optic cables (which a company called Made in Space has demonstrated can indeed be produced under zero-gravity conditions) could certainly qualify.

Finally, the sales volume of the product must be sufficiently high. If the basic costs of running the orbital lab are $40 million per year (a number that would allow shipping about one hundred tons per year of various supplies, assuming half the budget was so dedicated), the operation will still fail if only twenty tons of its $2,000/kg wholesale-priced product can be sold. To allow for all the business's costs, at least $80 million in gross revenue would be needed, or forty tons of product. Let's say that the end use for the product was a drug or computer chip selling retail for $200 for a hundred-gram unit. In that case, four hundred thousand units would have to be sold per year. Given a sufficiently desirable and unique product, sales numbers such as these are entirely feasible.

SPACE BUSINESS PARKS

The idea of the space business park is *not* to define the business but to create the infrastructure to support any need. If you build it, they will come—or so the theory goes. In other words, you build a large space-craft with a truss, a power array, attitude control systems, and some pressurized modules, and then you announce that you have space on orbit for rent. Perhaps your first customer might be an orbital research outfit. That would be logical; as we have seen, of all the orbital busi-nesses we have discussed so far, research has the best chance of pro-ducing a big profit under near-term technological assumptions. In the course of its investigations, the research company may discover a unique product that, contrary to their initial hopes, can only be pro-duced on orbit. In so doing, they are forced to rent additional modules from you for factory space. If necessary, you expand your space busi-ness park with additional pressurized modules to meet this demand. The operation of the orbital factory might create sufficient demand to drop the cost of providing orbital accommodations, thereby improving the economics of space tourism. At *that* point, the space hotel entre-preneur will find the needed funding, and you add on the deluxe bedroom modules with the reflecting mirror walls.

The space business park scheme has the advantage of being evo-lutionary, with an initial form of space-based activity (research) that could be viable even under current launch prices and which, moreover, will have a foundation of experience in the operation of the International Space Station, but with many advantages over that predecessor. For example, your business park would certainly eliminate the ISS's Kafkaesque process for permitting activities. Moreover, even if the ISS had subsidized its lab rental to the point where it was free, private pharmaceutical companies would generally prefer to pay premium prices for orbital research space in a place where they could keep the results of their investigations secret.

The lack of definition inherent in the concept of the mixed-use space business park is both its greatest strength and greatest weakness. It offers flexibility, which allows the business to avoid being trapped by fixed ideas of space enterprise that may prove to be crackpot. But its business plan had better be based on more than a "If we build it, they will come" philosophy. A solid set of advance commitments from well-funded orbital research projects or other initial customers will likely be required to get the business park off the ground. Fortunately, as transportation costs drop, such commitments will become increasingly obtainable.

BRINGING THE WORLD TOGETHER

For the past two decades, the volume and speed of global wireless communication has been expanding at an astonishing rate, with both increasing tenfold every five years.[5] In 2000, the world's total global data traffic was four thousand terabytes (TB) per month. It is now (2019) approaching forty million TB/month and by 2030 is projected to reach four billion TB/month.

In 2000, wireless data allowed a minority of the citizens of advanced sector countries to use mobile telephones. Today, a third of the world's population has smartphones, providing access to the internet at speeds enabling television transmission. By 2030, the data grid is expected to reach nearly every person on Earth, with sufficient power to control and coordinate the movements of billions of self-driving vehicles.

The market for data delivery is already gigantic, and it's exploding.

How can its enormous demand be met? And who is going to get paid to do the delivery?

Space entrepreneurs believe they have the answer. They will bring the world together with vast fleets of communications satellites. In the process, they will greatly enrich the world, while making themselves the richest people who have ever lived.

Current communication satellites operate in geosynchronous orbit, thirty-six thousand kilometers above the surface of the Earth. At this height, they orbit the Earth at exactly the same rate that it turns, thereby keeping themselves in the same place in the sky as seen from the ground. This feature, as the concept's inventor, science fiction writer Arthur C. Clarke, realized, is extremely convenient for communications purposes, as it allows ground-based antennas to be fixed in their aim, while permitting nearly the entire Earth to be covered by just three satellites spaced 120 degrees apart around the planet's equatorial plane. But to transmit from such a distance, the satellites need to be large and power hungry, and thus expensive. Moreover, it takes about 0.25 seconds for radio signals to make the seventy-two-thousand-kilometer round-trip from the Earth's surface up to geosynchronous orbit and down again. While this does not matter for applications like one-way radio or TV broadcasting, it poses serious problems for two-way communication. For long-distance telephone calls, the quarter-second signal time delay each way can be quite annoying. For systems attempting to remotely control machinery under dynamic conditions (for example, self-driving cars or aircraft), it could potentially be catastrophic.

If the satellites could orbit lower, say 1,200 kilometers, the time delay could be cut thirtyfold and the transmission power each way reduced a thousandfold. But satellites at that height orbit the Earth every two hours and can only been seen from the ground when they are relatively close by. As a result, they are constantly moving in and out of view, with each one only providing temporary coverage to a limited ground area as it travels along its orbital track. If you want to provide global communication coverage using spacecraft at such altitudes, you can't do it with three satellites; you will need *thousands* of them.

So that's the plan. And the race to pull it off is already on.

The first out of the box was a company called WorldVu, founded by a group of Google employees including Greg Wyler, Brian Holz, and David Bettinger.[6] Subsequently renamed OneWeb, in January 2015 the group obtained backing from the Virgin Group, Google, and Qualcomm to build and deploy a constellation of 648 small satellites, each with a mass of about two hundred kilograms, orbiting at an altitude of 1,200 kilometers. This would be a very large constellation. By comparison, the total number of satellites currently orbiting the Earth is about 1,500.

But the plan only got bigger. By February 2017, OneWeb announced that it had already sold out the capacity of its 648 satellites—before a single one had been launched or even built. So, raising another billion dollars from the SoftBank Group, the company expanded its planned constellation to 2,420 satellites.[7] Shortly thereafter, Samsung proclaimed it had plans for a Space Internet altitude constellation consisting of 4,600 satellites, orbiting at 1,557 kilometers altitude.[8]

Not to be outdone, in May 2017, Elon Musk announced that his SpaceX company would be fielding its own constellation, named Starlink, consisting of 4,425 small satellites orbiting at 1,200 kilometers, operating in eighty-three orbital planes. In turn, these craft would be supplemented with a further 7,518 satellites flying at just 340 kilometers altitude.[9] Once fully deployed, the constellation would cut time delays by a factor of one hundred while increasing data rates per unit power by a factor of ten thousand relative to current geosynchronous orbit communication systems. Test satellites were launched in 2018, with the first operational satellites to be deployed in 2019 and the full system operational by 2024.

These three constellations alone comprise some nineteen thousand satellites—more than ten times all existing spacecraft put together. The spacecraft they will employ are small and probably will be launched fifty or more at a time. Launching them will therefore require four hundred medium-lift (approximately ten tons to orbit) booster flights, which over an eight-year period would amount to about fifty extra launches per year. The income these satellites promise to generate is so great that they can be profitably deployed even at current launch costs. But as launch costs collapse, the economics are going to get even better. This will no doubt lead to further generations of constellations ever more advanced in their technology and fielded by contenders from all over the world.

Where this will all lead is impossible to say. But one thing is certain: the communications revolution has just begun.

THE CUBESAT REVOLUTION

Since the dawn of the space age, the cost to get satellites into orbit has been driven by the high launch price per kilogram of expendable boosters. One solution for this problem is the lower the freight rate by making the launchers reusable. There is another way, however, to solve this problem, and that is to slash satellite launch costs by cutting their weight. Miniaturization has made great strides in computers and other areas of electronics. Indeed, a good cell phone today packs more computing power than the room-sized mainframes that enabled the Apollo program. If we can shrink computers by a factor of a thousand, why not satellites? At $5,000 per kilogram, a ten-thousand-kilogram satellite would cost $50 million to launch, but a ten-kilogram unit only $50,000. If you want to get into space on the cheap, this could be the easiest way to do it.

Such is the approach taken by professors Robert Twigg (of Stanford University) and Jordi Puig-Suari (of Cal Poly, San Luis Obispo), the inventors of the CubeSat. By the late 1990s, it had become clear to these two that it should be possible to build a minimal-capability satellite with a mass of one kilogram and a volume of one liter, or a cube about four inches on a side. Even at current high launch costs, such units would be cheap enough to get into space on the kind of budgets available to universities, allowing students to get great hands-on experience by building real spacecraft and seeing how well they functioned in orbit. But fortunately, while such educational applications were seen as the purpose of the first CubeSats, the inventors developed the concept further, planning ahead to enable more capable satellites by putting together groups of cubes. Designers were quick to take advantage of this feature, creating

CubeSats composed of two, three, six, and twelve units (known as 2U, 3U, 6U, and 12U systems), whose capabilities rapidly increased as a result of the added units, continued advances in electronics, and the creation of a market for ultraminiature spacecraft instrumentation and propulsion systems. By 2005, ten CubeSats had been launched, by 2010 seventy, with the cumulative figure soaring to 420 by 2015. By the latter date, the CubeSats had become so capable that of the one hundred CubeSats launched in 2015, only five were school projects. The rest were serious spacecraft launched by NASA, the military, or commercial imaging and remote sensing companies.[10]

A favorite configuration is the 6U version, with dimensions of ten by twenty by thirty centimeters, or about the size of a shoebox. As of this writing, there were several companies offering to build these systems, complete with communications, attitude control, and foldout solar arrays, for less than $1 million. With a typical mass of six to ten kilograms, these little birds are very cheap to launch, yet are powerful enough to perform important scientific missions. In 2018, two of them, named MarCO-A and MarCO-B, flew along with NASA's InSight Mars lander as hitchhiker payloads and successfully fulfilled their mission of providing a real-time communication link for the primary spacecraft during its entry, descent, and landing on the Red Planet. Also, in that same year, NASA received a host of proposals from many organizations to send such vehicles to the lunar orbit to search for water ice or other resources on the surface of the moon. (My own company, Pioneer Astronautics, is one of the competitors. We are leading a team that has developed a proposed mission called RISE, for Radar Ice Satellite Explorer, which will consist of four 6U CubeSats that will go into lunar polar orbit and scan the entire moon using ground-penetrating radar as it turns beneath them, hopefully finding water ice near the poles and subsurface caverns elsewhere.) It is expected that there will be a number of winners, which will begin to take flight as soon as 2021.

Figure 2.3. The RISE mission would use four 6U CubeSats, each the size of a shoebox, to search the moon with ground-penetrating radar for polar ice and subsurface caverns. *Image courtesy of the RISE team.*

This is just the beginning. Once the space launch revolution is factored in, the cost of sending a highly capable ten-kilogram 6U CubeSat could drop from $50,000 to as low as $2,000, making space exploration an affordable activity not just to government, corporations, universities, and billionaires but even to private individuals of middle-class means.

Do you have a space exploration mission you would like to do yourself? Pretty soon, you may have your chance.

POWER FROM SPACE?

Many who see orbital commerce as the driving force for the development of a spacefaring civilization look to the generation of electricity for use on Earth by large solar power satellite (SPS) systems. Were it possible to generate electric power in space for terrestrial consumption at competitive rates, the market would be nearly unlimited. Vast numbers of huge SPS systems would then be built, and their construction and operation would require a huge fleet of reusable medium- and heavy-lift launch vehicles. Truly cheap access to space with booster systems of every payload capacity would be rapidly developed, and the doorway to the final frontier thrown wide open. According to advocates, such as Princeton professor Gerard O'Neill, such commerce could then provide the economic foundation for the development of large colonies, literally cities in space, in high Earth orbits, and this vision has served to motivate many space entrepreneurs, notably Blue Origin founder Jeff Bezos.

In space, solar energy is available twenty-four hours a day, not masked by the dulling effect of the Earth's atmosphere. Moreover, while most terrestrial solar arrays are fixed in orientation, an orbital solar array can track the sun. Avoiding the atmosphere increases the effective solar brightness by a factor of about 1.5, while the ability to track the sun multiplies the average power produced by the orbiting array by a factor of four. Thus, when both advantages are considered, an orbital solar array can produce a time-averaged output about six times greater per unit area than its counterpart fixed in orientation on the ground in an equatorial desert. The SPS unit beams its solar-produced power via microwaves to Earth, where it is received by a "rectenna." The microwave energy is then converted to high-voltage alternating power for consumption on the consumer grid. About half the power would be lost in the beaming process, reducing the orbital array's advantage over its terrestrial counterpart to a factor of three. However, as a countervailing advantage, the groundside rectenna is smaller than a solar array, and cheaper, and can be put nearly anywhere in the world, including places where solar power is frequently unavailable due to weather. So, once a SPS is in operation over the appropriate hemisphere, a relatively cheap rectenna could be installed nearly anywhere in that hemisphere to obtain power. This could make enormous quantities of electricity available in remote areas of the third world and avoid the need for installing expensive power generation equipment in countries where political instability might make such installations insecure.

But what would the price of such power be? At current launch costs and solar panel weights, the business case for SPS is totally hopeless. So let's consider instead a case assuming launch costs have been reduced to $200 per kilogram and solar cells cut to one-fifth their current mass-per-unit power. Solar panels with a mass/power ratio of twenty kilograms per kilowatt when operating at the Earth's distance from the sun are currently available. Panels five times lighter than these would therefore have a mass/power ratio of about 4 kg/kW. However, half the power is lost during transmission to Earth, and the weight of the SPS spacecraft, including all supporting structure, mechanisms, and attitude control systems and the microwave transmissions system, could be expected to be at least equal to the weight of the solar panels themselves. Thus, net delivered power produced by the SPS spacecraft would be closer to 16 kg/kW.

Now, the SPS spacecraft could not be in low Earth orbit. If it were, it would zip around the Earth once every ninety minutes and be unable to provide constant, or even frequent, service to a rectenna station on Earth. Instead, the SPS would have to be in a slow-moving high orbit, with the best choice being geosynchronous (GEO), thirty-six-thousand kilometers up. At that altitude, the SPS would orbit the Earth once every twenty-four hours, and since the Earth turns at the same rate, this would allow the satellite to hover over a fixed position on the Earth's equator. While the orbit of the SPS would be equatorial, its high altitude would give it a good line of sight for transmission over most of the hemisphere. The cost of delivering payloads to GEO, however, is about four times that of LEO, running in the range of $20,000/kg today, and would still be $800/kg after the space launch revolution. Therefore, the *launch cost* of the SPS would be about $12,800/kW, or $12.8 billion for a one-thousand-megawatt unit suitable for providing the power needs of a city the size of Denver. But that's just the SPS launch cost. If we add in the costs of assembly (the 1,000 MW SPS would be more than five square kilometers in size and would weigh eight thousand tons); maintenance; insurance; spacecraft hardware; hardware, construction, and real estate costs for the rectenna and its power conditioning system; salaries; taxes; and so on, the price of the total SPS would undoubtedly run at least $40 billion. That's an order of magnitude more expensive than any similar output unit of any kind— natural gas, nuclear, solar, wind, you name it—built on the ground. At these installation prices, the fact that the SPS requires no fuel would make very little difference. Just the interest on $40 billion would be about $2 billion per year! If we add in maintenance and depreciation over twenty years, the cost would be at least $6 billion per year. That boils down to a user price of $0.68/kWh (assuming that nothing is added for profit), more than *ten times* the $0.06/kWh currently prevailing in the United States.

This suggests that for SPS to become commercially competitive as a source of baseload power, not only would the price of space lift need to drop by a factor of twenty-five (to $200/kg to LEO), but the mass of solar power systems would *also* need to drop by a factor of fifty. That could be quite a stretch.

The key problem with the SPS concept as a source of baseload power is that while it is technically feasible, it attempts to provide a common commodity—electricity—that can be produced much more cheaply in any number of alternative mundane ways. In that sense, it is like a business plan to provide gravel for road construction by importing it from the moon, ignoring the fact that many cheaper sources are available locally. If you are going to import something to Earth from space for sale, it needs to have unique attributes, as could be the case for zero-gravity-produced pharmaceuticals.

What is the unique attribute of SPS power that could potentially justify its higher price? Fortunately, it has one: it can be instantly projected anywhere. So rather than consider SPS as a competitor to one-thousand-megawatt municipal nuclear or natural-gas-fired power plants providing cheap bulk power, it might be better to think of it as a technology for supplying power on demand to remote locations, replacing the one-to-ten-megawatt-class diesel generators that are currently used to provide power to off-grid Arctic oil drilling sites or isolated military bases at great expense, logistical difficulty, and, in some cases, serious risk. Such customers might well be happy to pay the higher SPS price to get the power they need.

Instead of one-thousand-megawatt SPS units with a mass of eight thousand tons trying to compete with cheap grid power, SPS designs should focus on five-megawatt systems weighing forty tons, providing premium power to off-grid customers who really need it and are willing to pay the price. Small-scale power grids in isolated developing-sector locations are also a potential market. Rapid provision of emergency backup power deliverable via field-deployable rectenna systems in the wake of disasters is another possibility. In the aftermath of Hurricane Maria in 2017, much of Puerto Rico was left without power for months, with serious consequences for public health, safety, and employment. Even at $0.68/kWh, SPS-provided electricity would have been quite welcome under such conditions.

Electricity can be worth a lot if you otherwise won't have it. It can be a lifesaver. While SPS technology seems unlikely to provide humanity with any substantial fraction of its overall power needs for the foreseeable future, it could help ensure that no people, anywhere, are ever left completely powerless.

Such a niche role might seem like a rather small outcome relative to O'Neill's grand vision for solar power satellites. But all big things start out as little things, so who knows?

DETERRING WAR

The United States needs a new national security policy. For the first time in more than sixty years, we face the real possibility of a large-scale conventional war, and we are woefully unprepared.

Eastern and central Europe are now so weakly defended as to virtually invite invasion. The United States is not about to engage in nuclear war to defend any foreign country. So deterrence is dead, and with the German army cut from twelve divisions to three; the British gone from the continent; and American troops down to a thirty-thousand-person, nearly tankless remnant, the only serious and committed ground force that stands between Russia and the Rhine is the Polish army. It's not enough. Meanwhile, in Asia, the powerful growth of the Chinese economy promises that nation eventual overwhelming numerical force superiority in the region.

How can we restore the balance, creating a sufficiently powerful conventional force to deter aggressors? It won't be by matching potential adversaries tank for tank, division for division, replacement for replacement. Rather, we must seek to totally outgun them by obtaining a radical technological advantage. This can be done by achieving space supremacy.

To grasp the importance of space power some historical perspective is required. Wars are fought for control of territory. Yet for thousands of years, victory on land has frequently been determined by dominance at sea. In the twentieth century, victory on both land and sea almost invariably went to the power that controlled the air. In the twenty-first century, victory on land, sea, or in the air will go to the power that controls space.

The critical military importance of space has been obscured by the fact that, in the period since the United States has had space assets, all of our wars have been fought against minor powers that we could also have defeated without such advantages. Desert Storm has been called the first space war because the allied forces made extensive use of Global Positioning Systems (GPS). However, if we had possessed no such technology, the end result would have been just the same. This has given some the impression that space forces are just a frill to real military power—a useful and convenient frill, perhaps, but a frill nevertheless.

But consider how history might have changed had the Axis of World War II possessed reconnaissance satellites—merely one of many of today's space-based assets—without the Allies having a matching capability. The Battle of the Atlantic would have gone to the U-boats, as they would have had infallible intelligence on the location of every convoy. Cut off from oil and other supplies, Britain would have fallen. On the Eastern Front, every Soviet tank concentration would have been spotted in advance and wiped out by German air power, as would any surviving British ships or tanks in the Mediterranean and North Africa. In the Pacific, the Battle of Midway would have gone very much the other way, as the Japanese would not have wasted their first deadly air strike on the unsinkable island but sunk the American carriers instead. With these gone, the remaining cruisers and destroyers in Fletcher's fleet would have lacked air cover, and every one of them would then have been hunted down and sunk by unopposed and omniscient Japanese airpower. The same fate would have awaited any American ships that dared venture forth from the West Coast. Hawaii, Australia, and New Zealand would have subsequently fallen, and eventually China and India as well. With a monopoly of just one element of space power, the Axis would have won the war.

But modern space power involves far more than reconnaissance satellites. The use of space-based GPS can endow munitions with one hundred times greater accuracy, while space-based communications provide an unmatched capability of command and control of forces. Knock out the enemy's reconnaissance satellites, and they are effectively blind. Knock out their comsats, and they are deaf. Knock out their NAVSATs, and they lose their aim. In any serious future conventional conflict, even between opponents as mismatched as Imperial Japan was against the United States—or Poland (with one thousand tanks) currently is against Russia (with twelve thousand)—it is space power that will prove decisive.

Not only Europe but the defense of the entire free world hangs upon this matter. For the past seventy years, US Navy carrier task forces have controlled the world's oceans, first making and then keeping the Pax Americana, which has done so much to secure and advance the human condition over the postwar period. But should there ever be another major conflict, an adversary possessing the ability to locate and target those carriers from space would be able to wipe them out with the push of a button. For this reason, it is imperative that the United States possess space capabilities that are so robust as to not only assure our own ability to operate in and through space but be able to comprehensively deny this to others.

Space superiority means having better space assets than an opponent. *Space supremacy* means being able to assert a complete monopoly of such capabilities. The latter is what we must have. If we can gain space supremacy, then the capability of any American ally can be multiplied by orders of magnitude and, with the support of the similarly multiplied striking power of our own land- and sea-based air and missile forces, be made so formidable as to render any conventional attack unthinkable. On the other hand, should we fail to do so, we will remain so vulnerable as to increasingly invite aggression by ever-more-emboldened revanchist powers.

For this reason, both Russia and China have been developing and actively testing antisatellite (ASAT) systems. Up till now, the systems they have been testing have been ground launched, designed to orbit a few times and then collide with and destroy targets below one thousand kilometers altitude. This is sufficient to take out our reconnaissance satellites but not our GPS and communications satellites, which fly at twenty thousand and thirty-six thousand kilometers respectively. However, the means to reach these are straightforward, and, given their critical importance to us, there is every reason to believe that such development is well underway.[11]

The Obama administration sought to dissuade adversaries from developing ASATs by setting a good example and not working on them ourselves. This approach has failed. As a consequence, many defense policy makers are now advocating that we move aggressively to develop ASATs of our own. While more hardheaded than the previous policy, such an approach remains entirely inadequate to the situation.

The United States armed forces are far more dependent upon space assets than any potential opponent. Were both sides in a conflict able to destroy the space assets of the other, we would be the overwhelming loser by the exchange.

What we need are not ASATs but something much better: fighter satellites, fully analogous to fighter aircraft.

A fighter aircraft has two critical functions—it destroys enemy aircraft and protects our own. An ASAT, as generally conceived, only performs the first. As such, it is more analogous to an antiaircraft missile.

But the decisive weapon for achieving air supremacy has always been fighter aircraft. Only fighter aircraft can protect bombers and the rest from enemy fighters while also denying the enemy the use of the air. We need to do both in space.

We need fighter sats that can not only knock out adversarial space assets but patrol as escorts for our own reconnaissance, GPS, and communication satellites to protect them from enemy ASATs.

Such a fighter sat should be relatively small and cheap, perhaps on the order of a hundred kilograms, with an impulsive propulsion system that allows it to maneuver rapidly to intercept an approaching ASAT. It should also be armed with a standoff weapon system that allows it to destroy or deflect approaching ASATs at a distance. Perhaps a projectile weapon system firing small high-velocity rockets with limited guidance might be considered. If the fighter can maneuver well, such a short-range system might be satisfactory: even a hit that is sufficient to impact a small velocity change on an attacking ASAT could cause it to miss. Alternatively, directed energy systems or larger missiles with sophisticated guidance enabling interception at long distances could be employed. However, what is needed is a craft that transcends the ASAT's kamikaze mode of attack, because any defensive system based on such a principle could rapidly be overwhelmed.

The fighter sat's capability as an antisatellite system is also essential. After all, unless we can take out the adversary's reconnaissance satellites, the entire US Navy surface fleet will be readily visible and therefore extremely vulnerable to attack by enemy ICBMs, submarines, and other means. So, in the event of conflict, we will need to sweep enemy satellites from the skies. But this essential task cannot be trusted to mere ASATs because if the enemy creates fighter sats to protect their space assets, then our ASATs could be readily defeated.

Only American fighter sats capable of defeating the adversary's fighter sat escorts can be relied upon to get through.

Finally, it may be noted that fighter sats engaged with others will inevitably take losses. Therefore, for the defense to be effective, the ability to rapidly launch replacements must also be an essential part of the system architecture.

In the period before World War II, many airpower theorists argued that since bombers were the air assets that actually impacted operations on the ground, producing and operating large bomber fleets was the key to air superiority. This theory was wrong, and caused the Eighth Air Force, operating B-17s over Germany, to take horrendous losses while sharply limiting its effectiveness. It was only after long-range fighter escorts such as the P-51 Mustang were introduced that America was able to achieve the air supremacy necessary for victory.[12]

America's space power today is at risk because it is based on the same fallacy as that promoted by the prewar bomber theorists. Yes, it is true that the communications, GPS, and reconnaissance satellites comprise the business end of space power that actually influence the war below. It may not seem that they need protection because in all the years they have existed, we have not fought any wars against opponents with space power capabilities. But it is precisely against such more capable potential adversaries that they are most critically necessary.

Without our communication satellites, we would be unable to adequately coordinate our forces. Without our reconnaissance satellites, we would be nearly blind. Without our GPS system, the effective firepower of our forces would be reduced by orders of magnitude. Were we to lose such capabilities, our victory in any future conflict would be very much in doubt. Were an adversary able to eliminate our space assets while preserving their own, our defeat would be virtually guaranteed. Such an outcome is unacceptable.

We are not at war today, but the creation of the necessary systems will take some time. If we are to be sufficiently prepared to deter aggression, we need to start developing the essential capabilities now.

This battle for space supremacy is one the Western alliance can win. Because societies based on liberty are the most creative, no unfree potential adversary can match us in this area if we put our mind to it. We can and must develop ever-more-advanced satellite systems, fighter satellite systems, and truly robust space launch and logistics capabilities. Then the next time an aggressor commits an act of war against us or a country we are pledged to defend, instead of impotently threatening to limit its tourist visas, we can respond by taking out its satellites, effectively informing it in advance the certainty of defeat should it persist.

The entrepreneurial space launch revolution offers the free world the chance to obtain these critical advantages.

If Russia or China hopes to compete with such a Western military space initiative, there is only one way they could do it: they would need to also become free.

In which case, there will be no war.

THE LEAP BEYOND EARTH ORBIT

We are living on the brink of a new space age. Entrepreneurs have opened the way toward cheap access to space. As a result, all sorts of commercial activities in near-Earth space will soon become not only practical but profitable, and consequently, they will happen.

The invisible hand is unstoppable. If someone can make money doing something, it will be done, for good or ill, regardless of the wishes of kings, presidents, religions, or secret police.

That said, the invisible hand cannot be counted upon to do everything that needs to be done.

The private sector can be expected to fund the development of new and ever-more-advanced reusable launch vehicles to keep reducing the price of access to orbit and enable fast global travel. It may also finance the creation of extensive human operations on orbit, including, as we have discussed, orbital labs, manufacturing facilities, and space tourism. This is now possible because the technology, the method of operation, and a good deal of the market necessary for these launch and orbital operations have already been paid for by substantial government funding over the past four decades. There is nothing wrong with this; in fact, it follows a near-universal historical pattern of terrestrial frontiers being first opened either by governments or social groups motivated by transcendent purposes and only afterward developed by private commerce. As a result, the business case for a vast expansion of human activity in geocentric space has now been established.

But what of journeys to other worlds? People can be courageous, but money is timid; it prefers to reproduce itself in tried and proven ways. If your only fundamental goal is to make money, there are far more reliable ways to do so than to venture into the unknown. Thus, on Earth, developing new frontiers for profit has occurred only after such regions have been explored and pioneered at considerable risk and cost by individuals possessing rather different motives.

Government space initiatives over the past forty years have tamed near-Earth space to the point where it is now a potential arena for private enterprise. This is an extremely positive development. Global fast transport will provide the market for truly cheap reusable launch systems that will make orbital tourism a reality as well. Orbital labs, eventually supplemented by orbital manufacturing stations, will make available an array of products that may revolutionize medicine and computer technology. The combination of low-cost space access with orbital servicing operations will also allow the development of global communications systems whose capabilities will impact society in ways that exceed the imagination of most people today. For example, such augmented communication constellations could enable low-cost wristwatch-sized communication devices that would be able to access on a real-time interactive basis all the storehouses of human knowledge from anywhere in the world. In addition, they would enable their users to communicate very high volumes of data—including voice, video, and music—either to each other or to the system's central libraries. They will not only make possible the global coordination of billions of self-driving cars but provide the kind of previously impossible automated air traffic control that could make mass use of private aircraft (finally, flying cars!) a reality as well.

The practical value of such systems is obvious, but their implications go far beyond the practical into the social and historical. We will see nations thoroughly linked together, resulting in deep cultural fusions and a radical generalization of the dissemination of human knowledge. In a real sense, the establishment of the full range of worldwide communication and transportation services made possible by cheap space access represents the final step establishing humanity as a Type I global civilization.

That said, the fundamental problem facing the human race today—the creation of a true interplanetary spacefaring Type II civilization—will not be solved by developing orbital private enterprise in geocentric space. True, such operations will serve as a "school for sailors," training the people and honing the skills and organizations for future space ventures in a way analogous to how coastal fisheries helped to provide the sailors to handle the ships of the great nautical explorers of the past. They will also make many of the technologies needed to venture further much cheaper. But human beings will never settle Earth orbit, because there is nothing there to settle. We need to reach beyond. The entrepreneurs' help in providing low-cost transportation and other services to aid initial outposts growing into settlements will be vital. By creating many of the flight systems needed to go to the moon or Mars, they are sharply lowering the cost, risk, and schedule thresholds that stand in the way of a decision by political leaders to launch such a program. Ultimately, however, we need to decide as a society to take on the challenges of worlds beyond. And that is a trail to be blazed by those who live for hope and not for cash.

Chapter 3

HOW TO BUILD A LUNAR BASE

MOON DIRECT

The breakthrough success of the Falcon Heavy launch offers America an unprecedented opportunity to end the stagnation that has afflicted its human spaceflight program for decades. In short, the moon is now within reach.

Here's how the mission plan could work. The Falcon Heavy can lift sixty tons to low Earth orbit (LEO). Starting from that point, a hydrogen/oxygen rocket-propelled cargo lander could deliver ten tons of payload to the lunar surface.

We therefore proceed by sending two such landers to our planned base location. The best place for it would be at one of the poles, because there are spots at both of the moon's poles where sunlight is accessible all the time, as well as permanently shadowed craters nearby where water ice has accumulated. Such ice could be electrolyzed to make hydrogen/oxygen rocket propellant to fuel both Earth Return Vehicles as well as flying rocket vehicles that would provide the base's crew with exploratory access to most of the rest of the moon.

The first cargo lander carries a load of equipment, including a solar panel array, high-data-rate communication gear, a setup to beam microwave power with a range of one hundred kilometers, an electrolysis/refrigeration unit, two crew vehicles, a trailer, and a group of teleoperated robotic rovers. After landing, some of the rovers are used to set up the solar array and communication system, while others are used to scout out the landing area in detail, putting down radio beacons on the precise target locations for the landings to follow.

The second cargo lander brings out a ten-ton habitation module loaded with food, spare space suits, scientific equipment, tools, and other supplies. This will serve as the astronauts' house, laboratory, and workshop on the moon. Once it has landed, the rovers hook it up to the power supply, and all systems are checked out. This done, the rovers are redeployed to do detailed photographing of the base area and its surroundings. All this data is sent back to Earth to aid mission planners and the science and engineering support teams, ultimately forming the basis of a virtual reality program that will allow millions of members of the public to participate in the missions as well.

The base now being operational, it is time to send the first crew. A Falcon Heavy is used to deliver another cargo lander to orbit, whose payload consists of a fully fueled Lunar Excursion Vehicle (LEV). This craft consists of a two-ton cabin like that used by the Apollo-era Lunar Excursion Module, together with an eight-ton hydrogen/oxygen propulsion system capable of delivering it from the lunar surface to Earth orbit. A human-rated Falcon 9 rocket then lifts the crew in a Dragon capsule to LEO, where they transfer to the LEV. Then the cargo lander takes the LEV, with the crew aboard, to the moon, while the Dragon remains behind in LEO.

After landing at the moon base, the crew completes any necessary setup operations and begins exploration. A key goal will be to travel to a permanently shadowed crater and, making use of power beamed to them from the base, use telerobots to mine water ice. Hauling this treasure back to the base in their trailer, the astronauts will feed the water into the electrolysis/refrigeration unit, which will transform it into liquid hydrogen and oxygen. These products will then be stored in the empty tanks of the cargo landers for future use—primarily as rocket propellant but also as a power supply for fuel cells and a copious source of life support consumables.

Having spent a couple of months initiating such operations and engaging in additional forms of resource prospecting and scientific exploration, the astronauts will enter the LEV, take off, and return to Earth orbit. There they will be met by a Dragon—either the one that took them to orbit in the first place or another that has just been launched to lift the next crew—which will serve as their reentry capsule for the final leg of the journey back home.

Thus, until lunar hydrogen/oxygen propellant is available, each mission that follows will require just one $120 million Falcon Heavy and one $60 million Falcon 9 to accomplish. That's not too bad. But once the base is well established, there will be little reason not to extend surface stays to six months. Furthermore, with propellant available on the moon for the return trip, the crew could simply fly to the moon in the same LEV that had just returned the previous team. Mounting such a mission would only require lifting the new crew to orbit in a Dragon, along with six tons of propellant to refuel the LEV for a one-way flight back to the lunar base. This could be done with a single Falcon 9 (or Blue Origin New Glenn) launch. So assuming that the program's hardware purchases will roughly equal its launch costs, we should be able to create and sustain a permanently occupied lunar base at an ongoing yearly cost of less than $250 million. This is less than 1.3 percent of the space agency's current budget—a sum America or NASA (or Jeff Bezos, for that matter) can easily afford.

The astronauts will not be limited to exploring the local region around the base. Refueled with hydrogen and oxygen, the same LEV spacecraft used to travel to the moon and back can be used to fly from the base to anywhere else on the moon, land, provide on-site housing for an exploration sortie crew, and then return them to the base. We won't just be getting a local outpost: we'll be getting complete global access to an entire world.

NASA has not yet embraced this plan. Instead, it is proposing to build a lunar orbiting space station dubbed the Lunar Orbit Gateway. This boondoggle will cost several tens of billions of dollars, at the least, and serve no useful purpose whatsoever.[1] We do not need a lunar orbiting station to go to the moon. We do not need such a station to go to Mars. We do not need it to go to near-Earth asteroids. We do not need it at all. If we do waste our time and money building it, we won't go anywhere.

Americans want and deserve a human spaceflight program that actually goes someplace. If we want to get to the moon, we need to go to the moon. We now have it in our power to do so. Let's seize the time.

Establishing Humanity on the Moon

Earth's moon is a world with a surface area the size of Africa, somewhat justifying lunar colonization visionary Krafft Ehricke's designation of it as our "eighth continent."[2] As a destination for space colonization, the moon has the undeniable advantage that it is the closest of any major or minor planetary body, reachable with existing chemical propulsion in a three-day flight. It is also clear that we have the capability to establish permanent bases on the moon—after all, we had piloted lunar vehicles before we had VCRs, hand calculators, microwave ovens, or push-button telephones. The lunar surface contains vast amounts of oxygen, silicon, iron, titanium, magnesium, calcium and aluminum, tightly bound into rocks as oxides, but there nevertheless. Data substantiating the existence of these resources is given in table 3.1, which shows the results of chemical analysis of various Apollo lunar samples.[3]

TABLE 3.1. CHEMICAL ANALYSIS OF TYPICAL APOLLO LUNAR SAMPLES

Compound	Apollo 11 basalt	Apollo 14 breccia	Apollo 17 regolith
SiO_2	40.46	48.09	44.47
TiO_2	10.41	1.51	2.84
Al_2O_3	10.08	16.72	18.93
FeO	19.22	9.53	10.29
MgO	7.01	10.18	9.95
CaO	11.54	10.67	12.29
Na_2O	0.38	0.73	0.43

These resources give the moon an enormous advantage as a destination for colonization over geocentric orbital space, where there is nothing at all to work with. These materials could be used to produce part of the consumables, rocket propellants, power systems, and building or shielding materials to support lunar settlements or related activities. The moon also possesses scarce, but in principle obtainable, supplies of helium-3, an isotope otherwise naturally nonexistent in the inner solar system. Helium-3 offers a number of potentially important advantages as a fuel for thermonuclear fusion reactors and thus perhaps could provide a future lunar colony with a cash export commodity. The moon has a vacuum environment and a gravitational pull only one-sixth as large as the Earth, thus making it much easier to launch spacecraft from there than the home planet. This has caused many to speculate that the moon might serve as the optimal port of departure for interplanetary expeditions to points beyond. In addition, the airless environment of the moon provides a unique near-Earth location in which large-scale vacuum processing of various locally available materials can be undertaken. It also offers, as we shall see in chapter 9, a terrific platform from which to conduct astronomical research on the rest of the universe, with potential phenomenal benefits from the breakthroughs in physics that could result. These advantages have caused some to postulate ambitious plans for lunar settlement based upon commercial development.

There are problems, however. As an examination of table 3.1 indicates, while the moon's rocks and soils possess ample supplies of oxygen and several important metals, they are entirely lacking in such vital substances as organics, hydrates, carbonates, nitrates, sulfates, phosphates, and salts. The key primary biogenic elements of hydrogen, carbon, and nitrogen are present on the moon, but in general only in extremely rare quantities (about fifty parts per million) in surface materials impregnated by the solar wind. However, as a result of data returned by the Clementine (1994), Lunar Prospector (1998–1999). LCROSS (2009), and Lunar Reconnaissance Orbiter (2009–present) missions, there is now very strong evidence that large amounts of water ice have been condensed in permanently shadowed craters near the lunar poles.[4] As water can be electrolyzed into hydrogen and oxygen, an excellent rocket propellant combination, this is a huge plus for lunar development, provided we take proper advantage of it.

That said, the leading secondary biogenic elements, sulfur and phosphorus, are also on the rare side, with typical concentrations in lunar soils ranging from five hundred to one thousand parts per million. Among the leading secondary industrial elements, potassium, manganese, and chromium are reasonably common (about 2,000 ppm), but nickel and cobalt are rather scarce (about 200 ppm and 30 ppm, respectively, in typical regolith, though they may be more accessible in certain places as meteor impact debris), while copper, zinc, lead, fluorine, and chlorine are extremely hard to come by (5–10 ppm each). Helium is present in solar-wind-impregnated regolith in concentrations of about 10 ppm, while argon and neon can be found in concentrations of about 1 ppm each. The potentially commercially valuable helium-3, however, constitutes only one part in 2,500 of the total helium supply (the vast majority of which is ordinary helium-4), or just four parts per billion of the surface regolith. Nevertheless, the value of helium-3 is so high that studies indicate that it could be well worth mining, even at such low concentrations.

So, with all that in mind, how do we win the moon? I've already laid out the basic plan. Now let's discuss the details.

The Apollo program utilized the Lunar Orbit Rendezvous (LOR) plan, in which only a lightweight Lunar Excursion Module (LEM) went to the surface and back while its command module mother ship remained in orbit to reduce overall mission mass. However, if the objective is to establish a permanent lunar base, not just perform sorties to the moon, then the production of lunar hydrogen and oxygen propellants must perforce be an early base priority. Once lunar propellants are available, coming back from the moon via direct return from the surface offers a lower mass and much more economical approach to lunar exploration.

Furthermore, substituting direct return for LOR enhances overall program safety. If you are returning directly from the lunar surface, the launch window back to Earth is always open—there is no need to phase your flight plan with an orbiter. (The planned Lunar Orbit Gateway plan is even worse than the conventional LOR plan in this respect, since it is in a two-week orbit instead of the two-hour low lunar orbit employed during Apollo.) The risk of a failed LOR maneuver (which would cause loss of crew) is also eliminated. If you are using the LOR plan, you also face an unattractive choice regarding the maintenance of your required orbital spacecraft. You could decide to keep someone aboard the orbital craft watching over it, in which case you will subject a person to flight risk, zero-gravity health impacts, and radiation dose with no matching addition of your surface exploration capability. Alternatively, you could leave your orbital return ride unmanned, in which case it could be lost—stranding you and your friends on the moon—if a problem needing correction should develop while no one is minding the store.

If you are going to the moon to stay, rather than just briefly visit, the direct-return mission architecture is clearly the right way to fly. So that's why the Moon Direct plan employs direct flight to the lunar surface, and direct return to Earth orbit from the lunar surface, with no lunar orbit orbital rendezvous needed going either way.[5]

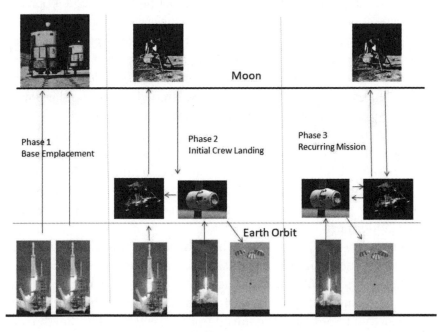

Figure 3.1. The Moon Direct program. In phase 1, two Falcon Heavy boosters are used to emplace base habitation modules and other cargo on the moon. In phase 2, one Falcon Heavy and one Falcon 9 are used to deliver the crew to the moon in a fueled LEV. In phase 3, only one Falcon 9 is used to deliver the crew to orbit and refuel the LEV. The crew then flies to the moon in the LEV, which refuels at the lunar base.

The base should be put at one of the poles because that is where water ice is to be found. Furthermore, while craters at the poles contain ice, nearby hills and mountains offer places that are nearly always illuminated by sunlight, thereby making solar power available without the two-week-long dark periods experienced nearly everywhere else on the lunar surface.

For example, Shackleton Crater, at the moon's south pole, is believed to contain millions of tons of water ice. Just one hundred kilometers away is an eight-kilometer-tall mountain called Mons Malapert (rising from three kilometers below the lunar reference elevation to five kilometers above) that is in full sunlight 87 percent of the time and partial sunlight 4 percent of the time. A base could be located between these two locations, with constant power provided by wired or wireless transmission from a solar array on Malapert's peak. Alternatively, there is a small crater just south of Mons Malapert which may contain ice. If so, the whole setup could be made much more compact. In either case, the proximity of round-the-clock solar power to water resources makes the polar location very compelling.

But before propellant production at the base is operational, any craft sent to the moon carrying crew for the setup operation needs to go there with all the propellant it needs to fly home. We therefore will need a cargo lander capable of delivering a crew capsule, together with its fully fueled ascent stage to the lunar surface. How big would such a lander need to be?

The Apollo-era LEM ascent vehicle had a dry mass of two metric tons.[6] The SpaceX Dragon capsule, which can carry crew and has a heat shield capable of withstanding reentry from Mars return (which is more demanding than the moon), has a mass of eight metric tons. The velocity increment ΔV required to fly directly back to Earth from the lunar surface and do a direct entry into the Earth's atmosphere is three kilometers per second. If a spacecraft is using hydrogen/oxygen rocket propulsion, which has an exhaust velocity of 4.5 km/s, it will require as much propellant as its own dry mass to execute such a maneuver (meaning that in technical terms, it would have a "mass ratio" of two, because its wet mass is twice its dry mass.) In this case, we will need enough propellant to send not only the Dragon but its ascent propulsion system, which might comprise an additional ton of tanks, engines, and avionics, on its way. So, all told, the Earth return system will have a mass of eighteen tons—eight for the Dragon, one for the rocket stage, nine for the propellant. We'll round it up to twenty tons to provide some margin. That's what our cargo lander would need to be able to deliver if we want to return using a Dragon.

In order to deliver twenty tons to the lunar surface, we would need to send forty tons to lunar orbit. As it happens, the 4.2 km/s ΔV required to send a spacecraft from LEO to lunar orbit is exactly the same as that needed to go from Earth orbit to Mars on a six-month transit trajectory. So, the same system that can send forty tons to lunar orbit can send an equal cargo on its way to Mars—which is enough to do the job. In the next chapter, I'll explain that part of the plan.

Unfortunately, sending forty tons to lunar orbit or trans-Mars injection is beyond the current capability of the Falcon Heavy or any other existing launch vehicle. SpaceX is working on creating such power, which can be obtained by refueling the Falcon Heavy second stage on orbit or by building a bigger booster, such as the planned Starship. But we don't have it yet. It could also be done by the NASA SLS, albeit at considerable expense (about $1 billion per launch), if it had a proper upper stage. But we don't have that yet either.

However, there is another option. Instead of using a Dragon, we can use a much lighter LEV modeled on the Apollo LEM for Earth return. Because it doesn't have a heat shield, the LEV will need to execute a 6 km/s ΔV to return—3 km/s for lunar ascent and 3 km/s for LEO capture. Using hydrogen/oxygen propellant, this requires a mass ratio of four, which is twice that needed by the Dragon return plan. But since the payload spacecraft is only one-quarter the mass, the total mass required is only half as great. So instead of needing a cargo lander capable of delivering twenty tons to the lunar surface, we need one that can land ten. This is well within the capability of the Falcon Heavy.

Thus the Moon Direct plan. First we land a couple of ten-ton habitation and supply modules, then we send a cargo lander to carry the crew out in a fully fueled LEV.

The crew needs to land near the habitation module. This should be straightforward: during Apollo we landed a crew within two hundred meters of a Surveyor probe landed a few years prior, and we have much better guidance systems today. After landing, the crew's first task will be to complete the deployment of the solar panels and get their pre-landed house in working order. Then it's off to the crater.

The crater is always dark and ultracold, with temperatures below forty kelvin. However, the crew vehicles, space suits, and teleoperated rovers will be equipped with lights and internal heaters, so they will be able to penetrate into the crater, explore, and gather ice, so long as they make sure to get out before their power is exhausted. Radioisotope power sources (either radioisotope thermoelectric generators or, better, because more efficient, dynamic isotope power sources) would therefore be ideal for such application, because they are essentially inexhaustible and, moreover, supply waste heat that could be used to help melt supercold ice. However, if nuclear systems are unavailable (which could be the case if the program is being conducted privately), a cable could be run into the crater from a photovoltaic station on its rim to power a recharging station inside.

But the best plan is to use beamed power. While transmitting electricity by microwave from geosynchronous orbit thirty-six thousand kilometers above the surface of the Earth to try to compete with ground-based commercial power is quite speculative, beaming power over a few dozen kilometers is a very different proposition. The size of a radio dish needed to transmit beamed power to a receiver of a given size scales as the transmission distance times the wavelength of the radio. So, transmitting the same wavelength radio thirty-six kilometers instead of thirty-six thousand only requires a transmitter dish one-thousandth the size—say, ten meters in diameter instead of ten kilometers—which makes everything much easier. Furthermore, on the moon we don't have to worry about radio wave absorption by the atmosphere.

As a result, we can employ millimeter-wavelength radio, an order of magnitude shorter than anything that will work on Earth. This will allow us to shrink the size of the required transmitter by at least an additional factor of ten. Good radio transmission through the Earth's atmosphere requires wavelengths higher than 1.5 centimeters to avoid water vapor absorption. On the moon, however, we could use 0.9-millimeter radio, greatly extending the range of a small power-beaming transmitter, despite the fact that it has a strong resonance with water. A twelve-meter transmitter dish set on the edge of the crater using the frequency would have a range of forty kilometers, transmitting to a portable two-meter dish that astronauts could use to reflect the radiation into a small, transparent tent placed over some deep-frozen icy regolith. Upon hitting the ground inside the tent, the microwaves would turn the ice into water vapor, which could then be piped into a condensing tank mounted on a trailer outside the tent.

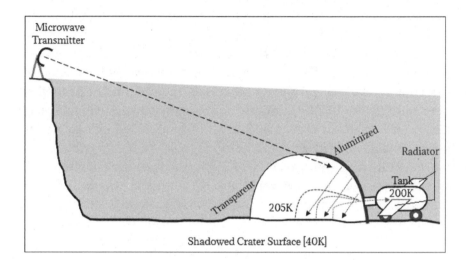

Figure 3.2. Concept for using beamed microwaves to extract water vapor from permafrost in a permanently shadowed lunar polar crater. Microwaves enter the tent from its transparent side and are then reflected down to send heat into the ground. Vapors that emerge are trapped in a trailer tank and then carted back to the base for electrolysis. *Image courtesy of Heather Rose, Pioneer Astronautics.*

Having mined a trailer tank's worth of ice, the team would bring it back to the base. The ice would then be warmed, turning it into liquid water. This would then be fed into the propellant production plant to make liquid hydrogen and oxygen, which could be stored in the now-empty propellant tanks of the cargo lander.

The first crew that reaches the moon will work toward reducing the process to practice, so that after their departure it can continue on an automated basis, teleoperated from Earth. Having accomplished that, they will take off directly for Earth in the LEV, leaving expansion of the power system and communication network and other further base development tasks to be done by the next crew. If desired, further unmanned habitation module/cargo flights can also be flown out to expand living space and bring in plenty of added equipment.

Once the propellant production is fully in gear and full-time power is available, crewed flights will no longer need to fly to the moon with their return propellant onboard. Instead, they will be able to travel to the moon by refueling a LEV that has just returned. Such a recurring mission could be enabled by a single Falcon 9 launch, which could lift the crew to orbit in a Dragon and supply the 7.5 tons of propellant needed to fly the LEV back to the moon. Alternatively, using a Falcon Heavy for launch, the crew could travel in an unfueled LEV carried by a cargo lander loaded with an additional eight tons of payload. With more gear arriving on every such flight, the base's capabilities will rapidly increase, its supportable population will grow, and mission durations will expand from weeks to months or even years. As this occurs, the base will transition from a local activity to a center supporting a vigorous, globe-spanning program of lunar exploration.

ACHIEVING LONG-RANGE MOBILITY

The moon is a world of rough terrain, with an extent the size of Africa. Such a world cannot be adequately explored on foot, or by ground vehicles. To get around the moon in any serious way, we are going to have to be able to fly. The moon, of course, has no air, so airplanes are out of the question. But by taking advantage of its polar ice to produce hydrogen/oxygen propellant, we will be able to fly all over the moon using rocket-powered ballistic flight vehicles.

The LEV weighs two tons. With a ΔV capability of 6 km/s, it would be able to fly from the lunar pole to a place 1,500 kilometers (fifty degrees of latitude) away, land, and return. As noted earlier, this would require a mass ratio of 4, or six tons of propellant for the round-trip. Alternatively, with a mass ratio of 2.5, it would be able to do 4 km/s, enough to take off from one pole and fly one-way to land at the other, using three tons of propellant for the trip.

Using such a system, a base at the south pole would give us access to much of the moon's southern hemisphere, one at the north pole would give us the north, and we could fly from one base to the other in less than an hour.

So the same LEV that allows us to go back and forth to the moon by lifting only six tons of propellant to orbit can also be used to enable lunar explorers operating out of a polar base to readily access most of their world.

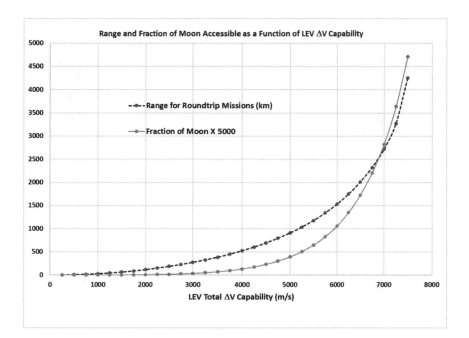

Figure 3.3. The range of LEV if used as a round-trip lunar excursion vehicle is shown dashed in black. The fraction of the total lunar surface made accessible is shown in gray (5,000 = 100 percent). For one-way flights, divide the ΔV shown by 2.

Producing one ton of liquid hydrogen/oxygen propellant from water ice requires about four hundred kilowatt-days of electricity. So, to make the six tons of propellant needed to power a long-distance sortie from the pole to forty degrees south and back would need 2,400 kW-days, or twenty-four days' activity by a one-hundred-kilowatt base power system.

The Loonies are going to be able to get around.

ENERGY FROM THE MOON

In recent years, two schemes have been proposed for supporting lunar colonization through the large-scale export of a cash commodity to Earth. In both cases, the product is energy. On the positive side, energy is an item with a guaranteed large and growing terrestrial market. On the negative side, it is a product without unique qualities. In the terrestrial electric power marketplace, a kilowatt is a kilowatt is a kilowatt, and consumers don't know or care whether the juice in their power lines comes from the moon or burning garbage at the city dump. Power can be sold in huge quantities, but only if it can be produced at or below the going rate. The two lunar-power-based development schemes embrace this challenge, but in radically different ways.

One of the concepts is to mass-produce solar power arrays on the moon out of lunar materials, deploy the arrays on the lunar surface, and beam the power to Earth.

There are a number of problems with this scheme. In the first place, it requires beaming energy across four hundred thousand kilometers of space. This could be a showstopper, because it means that both the transmitting antenna needs to be huge and the transmitter frequency needs to be very high, if the transmitted energy is to be focused on a receiver of reasonable size. For example, assuming a frequency of three gigahertz (ten-centimeter wavelength) on a ten-kilometer-diameter rectenna receiver on the ground on Earth, the system would need a lunar transmitting antenna forty kilometers in diameter. If the system were designed to operate at a higher frequency, say thirty gigahertz (one-centimeter wavelength), the transmitting antennas would scale down in proportion, but at the higher frequency, at least half the transmitted energy would be absorbed by Earth's atmosphere.

Producing all the silicon needed for large-scale solar arrays won't be easy either. As we can see in table 3.1, it's certainly true that SiO_2 is abundant on the lunar surface, but reducing this material to metallic silicon requires reacting it with carbon, which the moon essentially lacks. The required reaction, which must be done at very high temperatures (1,500°C), is:

$$SiO_2 + 2C \Rightarrow Si + 2CO \qquad (3.1)$$

While the carbon monoxide waste stream so generated can be processed to allow the carbon employed to be recycled and used again, in reality there is always loss in such cycling chemical engineering systems. So large supplies of hard-to-get carbon will be needed.

Another problem is that fixed on the moon's surface, the solar arrays would experience precisely the same average power generation losses due to day/night cycling (albeit at slower, longer intervals) and nonoptimal sun angles encountered by solar panels positioned on Earth's surface. Compared to Earth-based photovoltaics, the lunar array would only obtain the advantage of, at best, a twofold increase in power generation due to its perpetual clear weather. This advantage would be wiped out by the losses encountered in transmitting the power, even at low frequency where atmospheric losses are minimized. Thus, it is hard to see how, with all the additional costs involved in manufacturing and positioning the solar panels and their huge transmitting system on the moon, a lunar array could be economically competitive against Earth-based photovoltaics (which themselves have yet to become competitive against fossil fuels, hydroelectric, nuclear, wind, or geothermal power). Furthermore, since the Earth is always turning, it would be impossible for the lunar solar power station to provide continuous power to any one region. Instead, it would only be able to supply power for a few hours per day to any one place, making its utility somewhat problematic.

The other proposal, that of University of Wisconsin professors Jerry Kulcinski and John Santarius and Apollo astronaut Harrison Schmitt, is far more interesting.[7] These gentlemen propose to mine the lunar regolith for its helium-3 and then export this unique substance to Earth for consumption in terrestrial fusion reactors. Now, one obvious and frequently noted flaw in this plan is that fusion reactors do not exist. However, that fact is simply an artifact of the mistaken priorities of the ladies and gentlemen in Washington, DC, and similar places who have been controlling scientific research and development for the past few decades. While, driven by international rivalry, the world's national fusion programs did advance forcefully between 1960 and 1990, the decision to consolidate all of them into a unified global effort to build the International Tokamak Experimental Reactor (ITER) has caused nearly all progress to screech to a halt since the 1990s. Lack of funding and drive, not any insuperable technical barrier, has blocked the achievement of controlled fusion in the years since. The total budget for fusion research in the United States currently stands at about $400 million per year—in real dollars, about one-third of what it was in 1980—and no new major experimental machines have been built by the US Department of Energy for thirty years. Under these circumstances, the fact that the fusion program has continued to creep forward and now is approaching ignition is little short of remarkable.

But now, in large part as a result of SpaceX's demonstration that it is possible for a hard-driving entrepreneurial organization to achieve things that previously it was believed only the governments of major powers could attempt, several very promising and creative start-up companies have received substantial funding from venture capitalists. Just as with space launch, a very dynamic private fusion power race is now underway, giving us every reason to expect success.

All atomic nuclei are positively charged and therefore repel each other. In order to overcome this repulsion and get nuclei to fuse, they must therefore be made to move very fast while being held in a confined area where they will have a high probability of colliding at high speed. Superheating fusion fuel to temperatures of about 100,000,000°C gets the nuclei racing about at enormous speed. This is much too hot to confine the fuel using a solid chamber wall—any known or conceivable solid material would vaporize instantly if brought to such a temperature. However, at temperatures above 100,000°C, gases transition into a fourth state of matter, known as plasma, in which the electrons and nuclei of atoms move independently of each other. (In school, we are taught that there are three states of matter: solid, liquid, and gas. These dominate on Earth, where plasma exists only in transient form in flames and lightning. However, most matter in the universe is plasma, which constitutes the substance of the sun and all the stars.) Because the particles of plasma are electrically charged, their motion can be affected by magnetic fields. Thus, various kinds of magnetic traps (such as tokamaks, stellarators, and magnetic mirrors) have been designed that can contain fusion plasmas without ever letting them touch the chamber wall.

At least, that is how it is supposed to work in principle. In practice, all magnetic fusion confinement traps are leaky, allowing the plasma to gradually escape by diffusion. When the plasma particles escape, they quickly hit the wall and are cooled to its very low (by fusion standards) temperature, thereby causing the plasma to lose energy. However, if the plasma is producing energy through fusion reactions faster than it is losing it through leakage, it can keep itself hot and maintain itself as a standing, energy-producing fusion fire for as long as additional fuel is fed into the system. The denser a plasma is, the faster it will produce fusion reactions, while the longer individual particles remain trapped, the slower will be the rate of energy leakage. Thus, the critical parameter affecting the performance of fusion systems is the product of the plasma density (in particles per cubic meter) and the average particle confinement time (in seconds) achieved in a given machine. The progress that the world's fusion programs have had in raising this parameter, known as the Lawson parameter, is remarkable. Over the past forty years, it has been raised by a factor of ten thousand to reach the 10^{20} s/m^3 required for "breakeven," a condition in which the power produced by the reaction is equal to the external power being used to heat the plasma. If we can increase it another factor of three, we will reach ignition, in which, once started, the plasma produces enough energy to heat itself. At that point, the energy return of the system becomes unlimited.

Fusion can certainly be developed, and when it is, it will eliminate the specter of energy shortages for millennia to come. However, not all fusion reactions are created equal.

Currently, the world's fusion programs are focused on achieving the easiest fusion reaction, that between deuterium (hydrogen with a nucleus consisting of one proton and one neutron) and tritium (hydrogen with a nucleus consisting of one proton and two neutrons). Deuterium is nonradioactive and occurs naturally on Earth as one atom in six thousand ordinary hydrogens. It's expensive—about $10,000 per kilogram—but since an enormous amount of energy (about $5 million/kg worth, at current prices) is released when it burns, this is not really a problem. Tritium is mildly radioactive, with a half-life of 12.33 years, so it has to be manufactured. In a deuterium-tritium (D-T) fusion reactor, this would be accomplished by first reacting the fusion fuel as follows:

$$D + T \Rightarrow He4 + n \qquad\qquad (3.2)$$

Reaction 3.2 yields 17.6 million electron volts (MeV) of energy, about a million times that of a typical chemical reaction. Of the total yield, 14.1 MeV is with the neutron (denoted by "n"), and 3.5 MeV is with the helium nucleus. The helium nucleus is a charged particle, and so it is confined in the device's magnetic field, and as it collides with the surrounding deuterium and tritium particles, its energy will heat the plasma. The neutron, however, is uncharged. Unaffected by the magnetic confinement field, it will zip right out of the reaction chamber and crash into the reactor's first wall, damaging the wall's metal structure somewhat in the process, and then plow on until it is eventually trapped in a "blanket" of solid material positioned behind the wall. The blanket will thus capture most of the neutron's energy, and in the process, it will be heated to several hundred degrees Celsius. At this temperature, it can act as a heat source for high-temperature steam pipes, which can then be routed to a turbine to produce electricity. The blanket itself is loaded with lithium, which has the capacity to absorb the neutron, producing helium and a tritium nucleus or two in the process. The tritium so produced can be later separated out of the blanket materials and used to fuel the reactor. Thus, a D-T reactor can breed its own fuel.

However, not all of the neutrons will be absorbed by the lithium. Some will be absorbed by the steel or other structural elements composing the reactor's first wall, blanket cooling pipes, and other components. In the process, the reactor's metal structure will become radioactive. Thus, while the D-T fusion reaction itself produces no radioactive wastes, radioactive materials are generated in the reactor metal structure by neutron absorption. Depending upon the alloys chosen for the reactor structure, a D-T fusion reactor would thus generate about 0.1 to 1 percent of the radioactive waste as a nuclear fission reactor producing the same amount of power. Fusion advocates can point to this as a big improvement over fission, and it is. But the question of whether this will be good enough to satisfy today's and tomorrow's environmental lobbies remains open.

Another problem caused by the D-T reactor's neutron release is the damage caused to the reactor's first wall by the fast-flying neutrons. This damage will accumulate over time and probably make it necessary to replace the system's first wall every five to ten years. Since the first wall will be radioactive, this is likely to be an expensive and time-consuming operation, one that will impose a significant negative impact on the economics of fusion power.

So, the key to realizing the promise of cheap and radwaste-free fusion is to find an alternative to the D-T reaction, one that does not produce neutrons. Such an alternative is potentially offered by the reaction of deuterium with helium-3. This occurs as follows:

$$D + He3 \Rightarrow He4 + H1 \tag{3.3}$$

This reaction produces about 18 MeV of energy and no neutrons. This means that in a D-He3 reactor, virtually no radioactive steel is generated, and the first wall will last much longer, since it will be almost free from neutron bombardment. (I say "virtually no" and "almost free" because even in a D-He3 reactor, some side D-D reactions will occur between deuteriums that will produce a few neutrons.) In addition, no lithium blanket or steam pipes are needed. Instead, the energy produced by the reactor, since it is all in the form of charged particles, could be converted directly to electricity by magnetohydrodynamic means at more than twice the efficiency possible in any steam turbine generator system.

There are two problems, however. In the first place, the D-He3 reaction is harder to ignite than D-T, requiring a Lawson parameter of about 1×10^{21} s/m^3 to ignite. That should not be fundamental; it just means that D-He3 machines need to be a little bigger or more efficient at confinement than D-T devices. If we can do one, then in a few more years, we can do the other. The bigger problem is that helium-3 does not exist on Earth. It does, however, exist on the moon.

The solar wind contains small quantities of helium-3 and over ages of geologic time has implanted the surface layers of the lunar regolith with about four parts per billion of this unique isotope.

Four parts per billion is not much—for almost any substance, this would represent much too low a concentration to be economic for any sort of industrial recovery. At 4 ppb, you need to process 250,000 tons of raw material to obtain one kilogram of product. It certainly would not be worthwhile trying to refine gold from such dilute feedstock. But helium-3 is much more valuable than gold. A kilogram of gold, at today's prices, is worth about $40,000. A kilogram of helium-3, on the other hand, if burned in a fusion reactor using a 60 percent efficient MHD conversion system, would produce one hundred million kilowatt-hours of electricity. At a typical US current rate of $0.06/kWh, this represents a gross product value of $6 million/kg. This means that if a utility were willing to spend one-third of its gross revenue on fuel, helium-3 could sell for about $2 million/kg, or fifty times the price of gold. This is so high that even at current space transportation rates, it would be economical to ship such material from the moon.

Not only that, but there are other markets where electricity sells for a higher price than in the United States. Many Europeans pay four times as much as urban Americans for electricity, and there are other places where the going rate is considerably more. Indeed, the cost of power *in space* is currently about *a thousand times* the current US average, and while this will certainly come down with launch costs, for a long time to come, it's going to be quite pricey. So let's conservatively estimate a value for the He3 of at least $10 million per kilogram.

However, you still need to process 250,000 tons of raw lunar regolith to get it. That would mean "farming" an area one kilometer on a side and ten centimeters deep, then taking all that material to a "shake and bake" system where the soil would be heated to about 700°C, causing the helium-3 and all other embedded volatiles (including the several thousand times more common nonprecious helium-4) to outgas. If the processor could handle five tons a minute, this would take about thirty-five days, working round the clock. Isotope separation would then need to be performed to divide the 1 kilogram of helium-3 from the 2,500 kilograms of common helium-4. This would be done by bringing the helium to very low temperatures where it will fractionate, since the different isotopes have different boiling points.

In the process of producing the kilogram of helium-3, enough soil would be processed to also yield about ten tons each of solar-wind-implanted nitrogen, hydrogen, and carbon. While a thin basis for settlement, these by-products could go a long way toward relieving the logistics needs of a lunar mining base. In doing so, they would provide extra income to the helium-3 mining operation, since with $200/kg launch prices, it would cost about $1 million/ton to deliver payloads to the moon, yielding perhaps $15 million for the twenty tons of carbon and nitrogen (hydrogen will be available cheaper from lunar ice), assuming you could capture and sell most of it.

Processing 250,000 tons of moon dirt to make $25 million of cash product ($100/ton) may seem like a hard way to earn a living. Indeed, unless the operation is heavily automated, it is unlikely to be economical. However, because of the fundamental simplicity of the "shake and bake" processing system, such automation might be possible. One can envision groups of remotely operated bulldozers rapidly plowing tons of regolith onto a continuously moving conveyor belt, which dumps the material into an oven. Two ovens would be employed at a given site: one sealed for bakeout, the other one open for filling. When the first oven had caused its soil to outgas its volatile content, a trapdoor would open, and the "dried up" waste dirt would be dumped out the back onto a waste conveyor, while the front would open for a refill of fresh soil. Meanwhile, the second oven would seal itself up and start baking out its load of regolith. And so it would go, with each oven alternating roles in turn. Once the gases have been separated from the soil, they can be handled by fairly standard sorts of chemical engineering fluid-processing techniques that are highly susceptible to automation.

But there are other problems. The equipment required to process five tons per minute of lunar soil would have nonnegligible mass. An optimistic low-end guess for the ovens, the isotope separation system, the conveyor belts, and a small fleet of bulldozers and trucks would be on the order of two hundred tons. At launch costs of $5,000/kg to LEO, it would cost $1 billion to launch two hundred tons to orbit, but transportation to the moon using hydrogen/oxygen rocket propulsion increases the total amount to be lifted by a factor of five. Thus it would cost $5 billion to ship enough equipment to the moon to produce about $25 million per month or $300 million annual revenue. At this rate, it would take about sixteen years for the equipment to pay back its shipping cost, which is unattractive given the risks. But as the new generation of entrepreneurial space launch companies are beginning to show, in the long term it is within the realm of engineering feasibility to reduce launch costs by a factor of twenty-five relative to current rates. If such $200/kg-to-LEO launch prices were available, the cost of shipping the helium-3 mining gear could be recovered in less than a year, which would make the business potentially quite attractive, provided the cost of the equipment itself were not too great.

So, unlike the space solar power beaming schemes, the helium-3 business plan might actually be workable, but not immediately. It requires the development of economical controlled fusion power plants, advanced robotic mining and separating equipment, and very cheap space launch and translunar transportation systems. The uncertainties in all aspects of the plan make it clear that no one is going to invest big bucks to create a spacefaring civilization in order to be able to mine lunar helium-3 anytime soon. Rather, lunar helium-3 will become available as a resource to humanity as a result of developing a mature spacefaring civilization for other reasons. Helium-3 won't provide the magnet that will draw us into space, but mastery of space will give us helium-3.

This is important for two reasons. In the first place, lunar helium-3 represents a large resource, enough to power human civilization at its current level of energy consumption for about a thousand years. Second, and more important, however, D-He3 fusion represents not simply another source of energy but a new kind of energy. Aside from antimatter, which does not exist naturally in the habitable portions of the universe, D-He3 fuel has the highest energy/mass ratio of any substance known. If used as the fuel for a fusion rocket, the D-He3 reaction could produce exhaust velocities as high as 5 percent of the speed of light. Since rockets can generally be designed to achieve speeds of about twice their exhaust velocities, this means that a D-He3 rocket could reach 10 percent of light speed, making travel to nearby stars possible on timescales of four to six decades. This may seem long, but it is less than a human lifetime and, in any case, compares quite favorably with the millennia required for star flight using more conventional systems.

A tremendous amount of engineering advance will be required

before such high-performance fusion rockets become a practical reality. The point, however, is that the D-He3 rocket is one of very few systems based on currently known physics that is capable of enabling interstellar flight. That is a promise not to be taken lightly. It is the reason controlled fusion must be pursued despite all apparent obstacles, setbacks, and attractive alternatives. The stars are worth more than kilowatts.

A STEPPING-STONE TO THE COSMOS

The discovery of water on the moon gives new life to an idea that has been discussed in both science fiction and the astronautical engineering literature for some time: that of using a lunar base as a staging point for missions to worlds beyond. The idea is that since the moon has only one-sixth Earth's gravity and no atmosphere, it's possible to reach any destination in space much more easily from the moon than it is from the Earth's surface. Thus, if indeed rocket propellant can be made available on the lunar surface, the moon could well turn into an excellent refueling station and port of call for interplanetary traffic. This proposal was advanced even before water was detected on the moon—it was always known that the moon contained plenty of oxygen in the form of metal oxides in its rocks, and techniques have been demonstrated for getting it out. In particular, the mineral ilmenite ($FeTiO_3$), which occurs in about 10 percent concentrations in some lunar soils, can be reduced by hitting it with hydrogen at 800°C. The reaction involved is:

$$FeTiO_3 + H_2 \Rightarrow Fe + TiO_2 + H_2O \qquad (3.4)$$

The water produced is then electrolyzed to produce hydrogen, which is recycled back into the reactor, and oxygen, which is one of the net useful products of the system. (The other is iron, a very useful structural material that is produced by reaction 3.4 itself.) The high temperatures required, the necessity for hydrogen feedstock to replace leakage, and the need to mine and refine ilmenite make this system somewhat difficult. However, the reactor itself has worked on test stands at the Carbotek Corporation and at my own company, Pioneer Astronautics, and the balance of the system's complexities should be manageable at a mature lunar base.[8] Of course, the only useful propellant product is oxygen—if this were the best the lunar base could do, visiting spacecraft would still need to bring their own fuel (hydrogen, methane, or kerosene) to burn in the oxygen. But since for rocket vehicles, the oxygen generally constitutes at least 75 percent of the total fuel/oxygen propellant combination, a supply of lunar oxygen alone could still be quite useful. This is especially true for rockets operating from nonpolar bases. But where lunar water is available, then both oxygen and hydrogen can be provided, and the chemical process required to produce them becomes much simpler (only electrolysis is required) as well.

So, the idea of the moon as a refueling station is interesting. It has its possibilities, but also its limitations. Using lunar propellants as a means of refueling moon base spacecraft for their return to Earth or for hopping around the moon makes perfect sense. Surprisingly, however, using a moon base to refuel spacecraft on their way from Earth to Mars offers no benefits at all. This is because the spacecraft, its equipment, and the large majority of its provisions must come from Earth, and the rocket ΔV required to go from LEO to Mars (4.2 km/s) is less than that required to go from LEO to the surface of the moon (6 km/s). So even if there were rocket propellant, already made, sitting at a lunar base right now and available for free, it would make no sense for a Mars-bound spacecraft to fly there to get it! It would be easier and cheaper to fly to Mars directly. If the moon base operated a reusable lunar-surface-to-orbit shuttle and thus were able to provide propellant not only on the lunar surface but in lunar orbit, the situation would change, but not by enough. The ΔV to go from LEO to lunar orbit is 4.2 km/s, the same as that required to fly from LEO to Mars, so lunar orbital refueling of Mars-bound ships would still be pointless, especially since lunar-produced propellants delivered to lunar orbit would most certainly not be free.

However, once we start going well beyond Mars, the balance of benefits shifts. For example, the ΔV to go from LEO to Ceres, a planetoid in the heart of the main asteroid belt, is 9.6 km/s, which is greater than that required to go to either lunar orbit or the lunar surface. So, if lunar propellants could be made available in these locations cheaply enough (compared to simply lifting the required 9.6 km/s worth of propellant from Earth to LEO), moon-based refueling could be advantageous. As the destination chosen is moved further out, the required mission ΔV grows, and so do the potential benefits offered by lunar refueling. Of course, it would never pay to set up a lunar refueling station for the benefit of one or two outer solar system missions; the basic infrastructure would cost too much. But if there were regular interplanetary traffic, say to support mining operations in the main asteroid belt or among the gas giants of the outer solar system (where immense resources of precious metals and fusion fuel are to be found, as we shall discuss in chapters 5 and 6), a lunar refueling station could provide a vital supporting role.

The moon is not a stepping-stone to Mars but, rather, an alternative destination that can, and should, be pursued in parallel. But it could be a stepping-stone to the universe.

THE TRANSORBITAL RAILROAD

In this book so far, I've limited my discussion of the means of space travel to rockets. This is because they are the only technology that is currently practical. But there could be alternatives that are far more efficient, and the moon could be their birthplace.

But let me begin at the beginning.

Back in 1960, the Soviet newspaper *Komsomolskaya pravda* published an interview with engineer Y. N. Artsutanov containing a description of a novel means of Earth-to-orbit transportation.[9] In the Artsutanov scheme, a satellite placed in geostationary orbit would simultaneously extend cables down toward the Earth and in the opposite direction, keeping its center of mass, and thus its orbit, constant. This procedure would continue some thirty-six thousand kilometers until the lower cable reached the surface of the Earth, where it could be anchored and used to support elevator cabs. These cabs, in turn, could then be used to transport payloads up to the satellite, where they could then be released into geostationary orbit. If the payloads continued farther out along the cable, they would have greater than orbital velocity and could be released on trajectories that would take them to the moon, Mars, Jupiter, or beyond. In other words, Artsutanov had designed a skyhook, a virtual railroad to space.

This concept was published in both Russian and English, only to be widely ignored and promptly forgotten. It was eventually rediscovered in 1975 by Jerome Pearson of Wright-Patterson Air Force Base.[10] Pearson published a series of papers on the concept going into far greater detail than earlier authors, including derivations for system mass, tapered tether designs, and allowable rates for moving payloads along the tether without exciting dangerous vibrational modes. Subsequently, the geostationary tether concept was widely publicized by Arthur C. Clarke, who made it a central feature of his novel *The Fountains of Paradise*.

The skyhook concept as envisaged by Artsutanov, Pearson, and Clarke was a wonderful idea that offered a complete and easy solution to the problem of cheap Earth-to-space transportation. It had just one problem: It was impossible. It was impossible because if one places a load at the bottom of a geostationary tether, the bit of tether holding it must be thick enough to support that load. The next bit of tether must be thick enough to support not only the load but the bit of tether supporting the load. Thus, as it proceeds thirty-six thousand kilometers from the ground to geostationary orbit, the tether must get thicker and thicker, and its diameter and weight will grow exponentially. Depending upon the strength-to-weight ratio of the tether material assumed, the cross-sectional area of the tether at the satellite would be ten to twenty orders of magnitude greater than its area at its base, with similar gigantic ratios holding between the tether mass and the mass of the payload it is required to lift. Unless fantastical materials, such as thirty-six-thousand-kilometer-long single-crystal graphite fibers with incredible strength-to-weight ratios, were assumed, a tether designed to lift one ton would itself have to weigh quadrillions of tons. With real materials, the skyhook just wouldn't work.

Well, not on Earth, anyway. On the moon, things could be quite different. On the moon, the skyhook would only have to lift loads against one-sixth of Earth's gravity, and the length of the tether required could be a lot shorter too. The effects of these factors on the tether design equations are exponential, and the net result is that skyhooks really can be built on the moon using current state-of-the-art materials like Kevlar or Spectra. The lunar skyhooks would still have to weigh about one hundred times the mass of the payload they would lift or lower, but since they could be used again and again to transfer payloads back and forth between the lunar surface and orbit, an investment in such infrastructure could pay off well to support the operations of a mature lunar base.

There are several ways a lunar skyhook could be designed. One would be as a stationary system, with its center of mass at the "L1" Lagrange point in space, where the gravity of the Earth and the moon balance. Such a stationary skyhook, whose base would be dead center on the lunar equator on the side of the moon facing Earth, would need cable cars to deliver payloads up and down, just as in Artsutanov's original article. Another way, however, would be to position the center of mass of the skyhook in low equatorial lunar orbit and have the tether rotate at just the right speed so that its backward-moving tips have zero velocity with respect to the ground. Using such a system, you would just wait on the ground for the tether tip to come by, at which time you could step into the seat positioned at its tip and ride it up to orbit like a Ferris wheel! Six such "tether stops" could be positioned at different positions spread across the tether's ground track, and a tether tip would swing by each stop every couple of hours. If you wanted to travel rapidly across the moon, you could get on at one stop and get off at another. With an equatorial orbiting tether, there would only be six stops, and they would all be along the equator. If you put the rotating tether in a polar orbit, you could create a lot more stops scattered all over the moon, but service to any particular stop would be correspondingly less frequent.

But the idea I like best is to put the skyhook in a nonrotating polar orbit with its center of mass about 5,200 kilometers above the lunar surface. It would then orbit the moon with a period of about fourteen hours, with its bottom traveling at a velocity of just 0.22 km/s a few kilometers above the surface. An ascent vehicle (or an astronaut with a jet pack!) based at the pole would then use a very small ΔV to fly up, hover nearby, and catch it (0.22 km/s is five hundred miles per hour, about the speed of an airliner or a KC-135 aerial refueling tanker on Earth—jet fighters routinely rendezvous with tankers and refuel in the air at about 300 mph), after which it could simply hang on and descend and land, with a similarly tiny ΔV, at any latitude between its starting pole and the other. This would allow point-to-point travel all over the surface of the moon with hardly any propellant required! Of course, as the moon rotates beneath the orbiting tether system, locations would only get visited twice a month (except for the poles, which would be visited by each tether system once every fourteen hours), so several flying tethers would be needed to provide frequent, regular service everywhere.

But cheap global travel would only be one of the benefits. Instead of just hanging on to the bottom or the tether, the craft could be towed up by a cable car locomotive. It could then reach lunar orbit—or even go beyond it to high altitudes, from which it would have sufficient velocity to fly back to Earth, or outward to Mars or practically any other destination in the solar system, with almost no propellant required! Furthermore, a spacecraft coming from Earth could catch the tether station in orbit and ride it down, thus being able to land anywhere on the moon nearly propellant free, thereby doubling the amount of useful payload it could deliver to the surface. Because the tether would be relatively short (as skyhooks go), orbiting, and subject to lunar gravity rather than Earth gravity, its total mass would be less than twice what it could lift to orbit, paying for its investment in launch mass with just a few cargo deliveries.[11]

As noted, we can't use such a system to take payloads from the surface of the Earth into space. But we could use it to help send spacecraft from low Earth orbit to the moon, or beyond. For example, if we put the center of mass of a nonrotating tether spacecraft system at an Earth orbital altitude of 4,200 kilometers (a 6.14-km/s, three-hour orbit), a spacecraft starting in LEO in the same orbital plane could reach its bottom tip (hanging down at an altitude of 3,400 kilometers) by executing a ΔV of a mere 0.74 km/s. This small ΔV would put the spacecraft in an elliptical orbit with a perigee in LEO but an apogee at

3,400 kilometers, with the same 5.7 km/s velocity as the tether bottom tip, so there would be no ΔV required for it to latch on. A cable car locomotive could then take it and tow it up the line to 2,400 kilometers beyond the tether central station. At that 6,600-kilometer altitude, the tether would whip the spacecraft along at a velocity of 7.6 km/s, which, upon release, would give it the energy to take off on a translunar trajectory, a maneuver that without the help of the tether would take a ΔV of 3.2 km/s. If everything were phased correctly, it could catch the lunar orbiting tether and use it to land, thereby reducing the total trip ΔV from 6 km/s to as little as 1 km/s. This virtual transorbital railroad would cut the launch mass—or increase the payload—of any mission using it by at least a factor of four. Similar concepts could be used to enhance transport to Mars.

America's pioneers first went west on foot or by horse, raft, or canoe. But once towns were established at the far end of the trail, we took steps to greatly facilitate travel. We built turnpikes, and canals, and ultimately opened the way for all to move quickly and easily coast to coast with the Transcontinental Railroad. The first settlers of the moon and Mars will no doubt rough it at high cost and considerable discomfort in cramped little rockets. But their children or grandchildren will travel in style on the Transorbital Railroad.

Figure 3.4. On to Mars!

FOCUS SECTION: VIRTUAL REALITY

The spaceflight revolution will make it possible for millions of people to travel through space from point to point on Earth, and for thousands to people to travel across space to the moon, Mars, and eventually worlds beyond.

But with seven billion people on this planet, that still leaves the odds pretty high against you ever getting to experience an alien world.

Yet you should not lose hope. A new technology is rapidly advancing that could offer a partial solution. It's called virtual reality.

The way virtual reality works is that a vast data set is created that stores detailed views of a given environment from every angle. Then you put on a pair of goggles that give you that view, and associated equipment that measures your movements, and every way you look and move, you then experience that environment—albeit strictly through sight—just as if you were there. Or rather, you would experience it as if you were there in the character of a being like the protagonist in the movie *Ghost*, who could travel around and see everything but not touch or move anything. So admittedly it's not as good as being there. But it's the next best thing.

So let's say you want to go to the moon or Mars, but you can't afford the ticket price or take the required time off from work. If NASA, or SpaceX, or whoever is running the show is smart, they will have delivered a platoon of robotic rovers to the planet in question and tasked them with the job of photographing the entire region around the base for the purpose of creating a virtual reality simulation of the environment in question in a computer on Earth. Then you volunteer to help explore it. NASA might pay you for your time; Musk would probably charge you a small fee (he's sharper than they are). But either way, you get to go. You put on the goggles and start walking around on Mars, examining the rocks closely. If you see something anomalous—for example, a rock that looks more like a fossil than a mineral—you call over a rover, or an astronaut if one is around, and suggest that he, she, or it dust it off, hit it with a rock hammer, look at it with a microscope, or do whatever else might be appropriate to check it out.

Modern cameras can acquire images much faster than a small staff of professional scientists seeking to examine them can ever hope to match. NASA has already attempted to address this by inviting the public to look over photographs taken by its Mars orbiters, and significant discoveries have been made by amateur volunteers. Once we are on the surface of other worlds, far more data will be acquired than astronauts or Jet Propulsion Lab rover drivers will ever be able to explore on their own.

Millions of part-time ghosts will be needed to help them out. You can be one of them.

Chapter 4

MARS: OUR NEW WORLD

Beyond the moon lies Mars, the decisive step in humanity's outward migration into space. Mars is hundreds of times farther away than the moon, but it offers a much greater prize. Indeed, uniquely among the extraterrestrial bodies of our solar system, Mars is endowed with all the resources needed to support not only life but the development of a technological civilization. In contrast to the comparative desert of Earth's moon, Mars possesses oceans of water frozen into its soil as permafrost, as well as vast quantities of carbon, nitrogen, hydrogen, and oxygen, all in forms readily accessible to those clever enough to use them. Additionally, Mars has experienced the same sorts of volcanic and hydrologic processes that produced a multitude of mineral ores on Earth. Virtually every element of significant interest to industry is known to exist on the Red Planet.[1] With its twenty-four-hour day/night cycle and an atmosphere thick enough to shield its surface against solar flares, Mars is the only extraterrestrial planet that will readily allow large-scale greenhouses lit by natural sunlight.

Mars can be settled. For our generation and many that will follow, Mars is the New World.

MARS DIRECT

Some have said that a human mission to Mars is a venture for the far future, a task for "the next generation." Such a point of view has no basis in fact. On the contrary, as I explained in detail in my book *The Case for Mars*, the United States has in hand, today, all the technologies required for undertaking an aggressive, continuing program of human Mars exploration, with the first piloted mission reaching the Red Planet within a decade.[2] We do not need to build giant spaceships embodying futuristic technologies in order to go to Mars. We can reach the Red Planet with relatively small spacecraft launched directly to Mars by boosters embodying similar technology to that which carried astronauts to the moon half a century ago. The key to success comes from following a "travel light and live off the land" strategy similar to that which has served terrestrial explorers for centuries. The plan to approach the Red Planet in this way is called "Mars Direct."

Here's how the Mars Direct plan works. At an early launch opportunity, for example 2026, a single heavy-lift booster with a capability equal to that of the Saturn V used during the Apollo program is launched off Cape Canaveral and uses its upper stage to throw a forty-ton uncrewed payload onto a trajectory to Mars. Arriving at Mars eight months later, the spacecraft uses friction between its aeroshield and Mars's atmosphere to brake itself into orbit around the planet, and then lands with the help of a parachute followed by terminal deceleration rockets. This payload is the Earth Return Vehicle (ERV). It flies out to Mars with its two methane/oxygen–driven rocket propulsion stages unfueled. It also carries six tons of liquid hydrogen cargo, a one-hundred-kilowatt nuclear reactor mounted in the back of a methane/oxygen–driven light truck, a small set of compressors and automated chemical processing unit, and a few small scientific rovers.

As soon as the craft lands successfully, the truck is telerobotically driven a few hundred meters away from the site, and the reactor is deployed to provide power to the compressors and chemical processing unit. The hydrogen brought from Earth can be quickly reacted with the Martian atmosphere, which is 95 percent carbon dioxide gas (CO_2), to produce methane and water, thus eliminating the need for long-term storage of cryogenic hydrogen on the planet's surface. The methane so produced is liquefied and stored, while the water is electrolyzed to produce oxygen, which is stored, and hydrogen, which is recycled through the methanator. Ultimately, these two reactions (methanation and water electrolysis) produce twenty-four tons of methane and forty-eight tons of oxygen. Since this is not enough oxygen to burn the methane at its optimal mixture ratio, an additional thirty-six tons of oxygen are produced via direct dissociation of Martian CO_2. The entire process takes ten months, at the conclusion of which a total of 108 tons of methane/oxygen bipropellant will have been generated. This represents a leverage of 18:1 of Martian propellant compared to the Earth hydrogen needed to create it. Ninety-six tons of the bipropellant will be used to fuel the ERV, while twelve tons are available to support the use of high-powered, chemically fueled long-range ground vehicles. Large additional stockpiles of oxygen can also be produced, both for breathing and for turning into water by combination with hydrogen brought from Earth. Since water is 89 percent oxygen (by weight), and since the larger part of most foodstuffs is water, this greatly reduces the amount of life support consumables that need to be hauled from Earth.

The propellant production having been successfully completed, in 2028 two more boosters lift off the Cape and throw their forty-ton payloads toward Mars. One of the payloads is an unmanned fuel factory/ERV just like the one launched in 2026; the other is a habitation module carrying a crew of four, a mixture of whole food and dehydrated provisions sufficient for three years, and a pressurized methane/oxygen–powered ground rover. On the way out to Mars, artificial gravity can be provided to the crew by extending a tether between the habitat and the burnt-out booster upper stage and spinning the assembly. Upon arrival, the crewed craft drops the tether, aerobrakes, and lands at the 2026 landing site, where a fully fueled ERV and fully characterized and beaconed landing site await it. With the help of such navigational aids, the crew should be able to land right on the spot, but if the landing is off course by tens or even hundreds of kilometers, the crew can still achieve the surface rendezvous by driving over in their rover. If they are off by thousands of kilometers, the second ERV provides a backup. However, assuming the crew lands and makes rendezvous as planned at site number one, the second ERV lands several hundred kilometers away to start making propellant for the 2030 mission, which in turn flies out with an additional ERV to open up Mars landing site number three. Thus, every other year two heavy-lift boosters are launched—one to land a crew and the other to prepare a site for the next mission—for an average launch rate of just one booster per year to pursue a continuing program of Mars exploration. This would clearly be affordable. In effect, this "live off the land" approach removes the manned Mars mission from the realm of mega-fantasy and reduces it in practice to a task of comparable difficulty to that faced in launching the Apollo missions to the moon.

(See plate 5.)

The crew stays on the surface for 1.5 years, taking advantage of the mobility afforded by the high-powered, chemically driven ground vehicles to accomplish a great deal of surface exploration. With a twelve-ton surface fuel stockpile, they have the capability for more than twenty-four thousand kilometers' worth of traverse before they leave, giving them the kind of mobility necessary to conduct a serious search for evidence of past or present life on Mars—an investigation key to revealing whether life is a phenomenon unique to Earth or general throughout the universe. Since no one has been left in orbit, the entire crew has available to them the natural gravity and protection against cosmic rays and solar radiation afforded by the Martian environment, and thus there is not the strong driver for a quick return to Earth that plagued previous Mars mission plans based upon orbiting mother ships with small landing parties. At the conclusion of their stay, the crew returns to Earth in a direct flight from the Martian surface in the ERV. As the series of missions progresses, a string of small bases is left behind on the Martian surface, opening up broad stretches of territory to human cognizance.

Such is the basic Mars Direct plan. In 1990, when I first put it forward, it was viewed as too radical for NASA to consider seriously, but over the following years, with the encouragement of then NASA Associate Administrator for Exploration Mike Griffin and NASA Administrator Dan Goldin, the group at Johnson Space Center in charge of designing human Mars missions decided to take a good, hard look at it. They produced a detailed study of a Design Reference Mission based on the Mars Direct plan but scaled up about a factor of two in expedition size compared to the original concept. They then produced a cost estimate for what a Mars exploration program based upon this expanded Mars Direct would cost. Their result: $50 billion, with the estimate produced by the same costing group that assigned a $400 billion price tag to the traditional cumbersome orbital assembly of megaspacecraft approach to human Mars exploration that was embodied in NASA's 1989 "90 Day Report" (a report that caused sufficient congressional sticker shock to lead to the cancellation of President George H. W. Bush's Space Exploration Initiative). If scaled back to the original lean Mars Direct plan described here, the program could probably be accomplished for $20 billion. This is a sum that the United States, Europe, China, Russia, or Japan could easily afford. It's a small price to pay for a new world.

In essence, by taking advantage of the most obvious local resource available on Mars—its atmosphere—the plan allows us to accomplish a manned Mars mission with what amounts to a lunar-class transportation system. By eliminating any requirement to introduce a new order of technology and complexity of operations beyond those needed for lunar transportation to accomplish piloted Mars missions, the plan can reduce costs by an order of magnitude and advance the schedule for the human exploration of Mars by a generation.

MARS DIRECT 2020

The original Mars Direct plan was developed in 1989–1990. It is still sound. But in the three decades that have passed, two key developments have occurred that offer great advantages for its implementation.

The first, for which credit is due to the planetary exploration programs led by NASA and the European Space Agency, is the discovery of vast amounts of water on Mars. In 1996, NASA sent the Pathfinder lander and the Mars Global Surveyor (MGS) orbiter to Mars. Arriving in 1997, Pathfinder deployed Sojourner, the first Mars rover. This little craft proceeded to discover conglomerate rocks, proof of the past existence of liquid water on the Martian surface. Operating in Mars orbit for the following decade, MGS mapped thousands of water-created features on the Martian surface, including the very flat basin of what looks like a former northern ocean. Not only that, photos taken by MGS in 2000 and 2005 of the same crater show the appearance of what looks like the remains of a streambed running down its side between those two dates, indicating a modern-day transient flow of water. The only potential source for this would be a subsurface liquid water table. Then in 2001, NASA sent the Mars Odyssey orbiter to the Red Planet. Using a neutron spectrometer, which can detect hydrogen concentrations in the soil, Odyssey mapped the entire planet, finding water concentrations of at least 5–10 percent nearly everywhere, with some regions of continental size reporting as 60 percent water by weight. Then, in 2003, NASA sent the Spirit and Opportunity rovers to Mars, while the Europeans sent the Mars Express orbiter. Arriving in early 2004, the rovers found salt deposits left behind by water evaporation on the shores of ancient Martian lakes. Mars Express did even better. Using its MARSIS ground-penetrating radar, starting in 2008 it reported indications of subsurface water, leading up to an announcement in 2018 of its discovery of a twenty-kilometer-wide underground lake of *liquid* salt water near the Martian south pole.[3] Such a lake not only could serve as a resource for astronauts, it is an environment where ancient Martian life, having to retreat from the surface as it became cold and dry, could still survive. The possibility of such life was further reinforced by initially controversial detections of methane emissions by Mars Express, which were subsequently confirmed by NASA's Curiosity rover, which landed on Mars in 2012. Methane molecules have a limited lifetime in Mars's atmosphere because they are destroyed by ultraviolet light. As a result, any methane found today must have been made recently, and the only possibilities are either biological or hydrothermal—that is, either life itself or a place friendly to life. But the grand prize discovery was made by NASA's Mars Reconnaissance Orbiter (MRO), which, after arriving at the Red Planet in 2006, first used its superb optics to identify massive layered ice deposits exposed on the sides of steep terrain in subarctic latitudes, and then, in 2018, used its SHARAD ground-penetrating radar to identify hundreds of Martian glaciers, composed of pure or nearly pure water ice, covered by just a few meters of dust, at latitudes as low as 38 degrees north—the same as San Francisco or Athens, Greece, on Earth. The amount of water in these glaciers is remarkable. It is about 150 *trillion* cubic meters, which is enough to cover the entire Red Planet with a layer of water more than a meter deep![4]

What this means is that astronauts using the Mars Direct plan won't need to bring hydrogen to Mars to react with CO_2 to make their methane/oxygen rocket propellant. Instead, all the key ingredients for propellant making—or crop growth—or plastics or fabric manufacture—can readily be found on Mars.

The other key development is SpaceX.

In February 2018, after a six-year development effort, SpaceX finally flew its Falcon Heavy booster, a three-quarters-reusable system with a sixty-ton-to-orbit payload capability (about triple that of the space shuttle, the Lockheed Martin Atlas V, the Russian Proton, or the European Ariane V, and double that of the Boeing Delta IV heavy).[5] As noted in the introduction to this book, this was an extraordinary accomplishment, done in half the time for one-thirtieth the cost anticipated for such a project by mainstream aerospace managers. It is true that at sixty tons to orbit, rather than the approximately one hundred tons generally thought of as the capacity of a true heavy-lift launch vehicle, Falcon Heavy is a bit on the light side for sending humans to Mars. But it is good enough.

If a SpaceX Falcon Heavy launch vehicle were used to send payloads directly from Earth, it could land about twelve tons on Mars. Using a modified Mars Direct plan involving three launches per mission, we could start human Mars explorations with two-person crews in the very near future.

But if instead, the upper stage of the Falcon Heavy booster were refueled in low Earth orbit, it could be used to land as much as forty tons on Mars—that is, more than a Saturn V—and therefore more than enough to implement the full-scale Mars Direct plan. Furthermore, by designing the mission to stage off the Falcon Heavy second stage after a ΔV equivalent to translunar injection, the entire launch system would be reusable and able to support powerful payload deliveries to the moon as well.

(See plate 6.)

But Falcon Heavy is by no means the limit to SpaceX's ambitions.

In remarks at the International Astronautical Congress (IAC) in Guadalajara, Mexico, on September 29, 2016, SpaceX president Elon Musk revealed to great fanfare his company's plans for an Interplanetary Transport System (ITS).[6] According to Musk, the ITS would enable the colonization of Mars by the rapid delivery of a million people in groups of a hundred passengers per flight, as well as large-scale human exploration missions to other bodies, such as Jupiter's moon Europa.

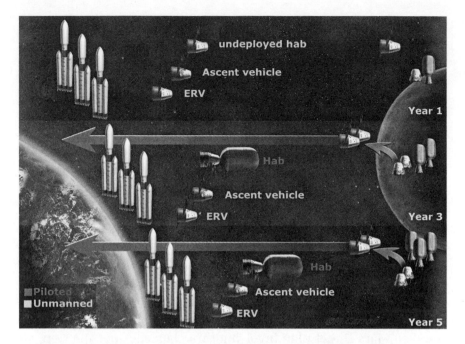

Figure 4.1. Mission sequence chart for the "Dragon Direct" plan. Every two years, three Falcon Heavies are launched, sending an ascent vehicle, an Earth Return Vehicle, and a piloted habitat. Inflatable hab modules are used to add ample living space to the Dragon capsules. *Image courtesy of Michael Carroll, Mars Society.*

I was among the thousands of people in the room (and many more watching online) when Musk gave his remarkable presentation, and I was struck by its many good and powerful ideas. However, Musk's plan assembled some of those good ideas in an extremely suboptimal way, making the proposed system impractical.

As described by Musk, the SpaceX ITS would consist of a very large two-stage fully reusable launch system, powered by methane/oxygen chemical bipropellant. The suborbital first stage would have four times the takeoff thrust of a Saturn V. The second stage, which reaches orbit, would have the thrust of a single Saturn V. Together, the two stages could deliver a maximum payload of 550 metric tons to low Earth orbit (LEO), about four times the capacity of the Saturn V.

At the top of the rocket, the spaceship itself—where some hundred passengers reside—would be inseparable from the second stage. Since the second-stage-plus-spaceship would have used its fuel in getting to orbit, it would need to refuel in orbit, filling up with about 1,950 tons of propellant (which means that each launch carrying passengers would require four additional launches to deliver the necessary propellant). Once filled up, the spaceship could head to Mars.

The duration of the journey would of course depend on where Earth and Mars were in their orbits; the shortest one-way trip would be around 80 days, and the longest would be around 180 days. (Musk stated that he thinks the architecture could be improved to reduce the trip to 60 or even 30 days.)

After landing on Mars and discharging its passengers, the ship would be refueled with methane/oxygen bipropellant made on the surface of Mars from Martian water and carbon dioxide, and then flown back to Earth orbit.

(See plate 7.)

The September 2016 ITS plan contained a number of notable strengths and weaknesses, which I explored in a lengthy review the following month.[7] On the strong side, the system was fully reusable, made maximum use of Martian resources by employing direct return from the surface using locally produced methane/oxygen propellant, and solved the problem of landing large payloads on Mars by employing the same breakthrough supersonic retro-propulsion technique that SpaceX has demonstrated on its Falcon rockets. On the weak side, the system was way too big, and flying such a giant rocket all the way to Mars and back would put huge unnecessary burdens on the propellant manufacturing systems on Mars—indeed, probably putting its power requirements beyond any nuclear or photovoltaic power systems likely to become available anytime soon. Furthermore, flying the ITS spaceship all the way to Mars and back would put the ship out of action for four years (since it would return too late to make the next launch window to Mars, two years after its own departure). A better plan would be to keep the ship in geocentric space and stage the trans-Mars payloads off of it, either from LEO (which would allow it to be used again the next day) or from a translunar injection (TLI) orbit (i.e., just short of Earth escape, which would allow it to be used again the next week).

Well, it would appear that SpaceX took my critique to heart, because the next year, when Musk presented his plan again at the IAC conference in Adelaide, Australia, he had scaled down the ITS (by now renamed the BFR) by a factor of almost four to a much more manageable 150 tons to orbit.[8] Furthermore, the new BFR (later renamed "Starship") had a "cargo variant" allowing it to carry payloads to LEO or TLI, which could then complete missions to the moon or Mars on their own.

Staged at LEO, with no need for on-orbit refueling, the Starship can send fifty tons on its way to the Red Planet—more than enough for Mars Direct. With orbital refueling, the Starship can lift 150 tons to TLI, from which it could fire off 120-ton cargos, with scores of Mars settlers riding the freight.

But best of all, now it's not just talk. As these lines are being written, the parts of the first Starship are being made. We are on our way.

Exploring Mars requires no miraculous new technologies, no orbiting spaceports, and no gigantic interplanetary space cruisers. We can establish our first small outpost on Mars within a decade. We and not some future generation can have the eternal honor of being the first pioneers of this new world for humanity. All that's needed is present-day technology, some nineteenth-century industrial chemistry, a solid dose of common sense, and a little bit of moxie.

MINI BFR

In the fall of 2018, Elon Musk let it be known that SpaceX was considering building a "Mini BFR" based on the upper stage of a Falcon 9 to use as a test article to support the BFR (subsequently renamed "Starship") development program. I was struck by the potential of such a system, so I wrote him the following memo.

Elon;

I saw that you are developing a Falcon 9 mini BFR for test purposes. That's great.

I hope you take it further and make it an operational system. This would offer many advantages.

1. It would give SpaceX a fully reusable medium-lift launch vehicle, the first in history, which could be a very profitable workhorse for the company.
2. It would provide a means of getting humans to Mars in the near term.

Consider;

3. A mini BFR spaceship, or a Small Falcon Spaceship (SFS) for short, could be refueled on orbit sufficiently for trans Mars injection with a single Falcon Heavy flight.
4. It would require an order of magnitude less propellant to fly back from Mars, and therefore an order of magnitude less surface power. This is very important. No one is going to give you bomb grade U235 to make a multimegawatt space nuclear reactor, and you will need to wait until well after you are dead to have NASA develop one. So if you are going to do propellant making on Mars, you are going to need to use solar power, and the size of the required system for the BFR could be a show stopper. But with an SFS you could probably do it.
5. Even when you have the BFR, and are using it for colonization, you would get much more action out of it by either working it as a fully reusable 150 t to LEO booster, or refueling it for TLI, and staging the payload off of it then, instead of sending the BFS all the way to Mars and back. That way you get to reuse it in days, instead of years. During colonization you are sending nearly all passengers one way, and only a few come back. The cheapest way for colonists to travel is to ride the freight, that is, fly out to Mars in hab modules that will stay there. You only need a much smaller vehicle to take a few returnees back. The SFS fits that bill, with much lower capital commitment and much lower power burden for the colony.
6. Unlike the BFS, the SFS already has a first stage that can lift it. This is a huge advantage. Why is Dragon superior to Orion? Because it can be launched. The same logic holds here.

Furthermore:

7. The ultimate market that could lead to cheap spaceflight is intercontinental travel. It will be a long time before you can fill a BFS with passengers at $20,000 a seat. You could fill an SFS much sooner.
8. For the same reasons that a SFS is superior to the BFS for Mars missions, it would also be superior for lunar missions.
9. You could do a mission launching a SFS out of LEO on a near Earth asteroid or lunar flyby mission in just a few years.
10. Creating an operating SFS system with provide you with plenty of heritage and operational experience for the design of the BFS.

Think about it.

All the best,
Robert

Let's hope he listens.

COLONIZING MARS

We hold it in our power to begin the world anew.
—Thomas Paine, 1776

The question of colonizing Mars is not fundamentally one of transportation. If we were to use the Starship or a comparable vehicle to launch habitats carrying settlers to Mars on one-way trips, firing them off at the same rate we launched the space shuttle when it was in its prime, we could populate Mars at a rate comparable to that which the British colonized North America in the 1600s—and at much lower expense relative to our resources. No, the problem of colonizing Mars is not moving large numbers to the Red Planet, but the ability to transform Martian materials into resources to support an expanding population once they are there. The technologies required to do this will be developed at the first Mars base, which will thus act as the beachhead for the wave of immigrants to follow. Initial Mars Direct exploration missions will approach Mars in a manner analogous to terrestrial hunter-gatherers and utilize only its most readily available resource, the atmosphere, to meet the basic needs of fuel and oxygen. In contrast, a permanently staffed base will approach Mars from the standpoint of agricultural and industrial society. It will develop techniques for extracting water out of the soil; for conducting increasingly large-scale greenhouse agriculture; for making ceramics, metals, glasses, and plastics out of local materials; and for constructing large pressurized structures for human habitation and industrial and agricultural activity.[9]

Over time, the base could transform itself into a small town. The high cost of transportation between Earth and Mars will put a strong financial incentive to find astronauts willing to extend their surface stay beyond the basic eighteen-month tour of duty to four years, six years, and more. Experiments have already been done showing that plants can be grown in greenhouses filled with CO_2 at Martian pressures—the Martian settlers will thus be able to set up large inflatable greenhouses to provide the food required to feed an expanding resident population. Mobile microwave units will be used to extract water from Mars's abundant permafrost, supporting such agriculture and making possible the manufacture of large amounts of brick and concrete, the key materials required for building large, pressurized structures. While the base will start as an interconnected network of Mars Direct–style "tuna can" habitats, by its second decade, the settlers could live in brick and concrete pressurized domains the size of shopping malls. Not too long afterward, the expanding local industrial activity will make possible a vast expansion in living space by manufacturing large supplies of high-strength plastics like Kevlar and Spectra that will allow the creation of inflatable domes encompassing sunlit pressurized areas up to one hundred meters in diameter. Each new reactor landed will add to the power supply, as will locally produced photovoltaic panels and solar thermal power systems. However, because Mars has been volcanically active in the geologically recent past, it is also highly probable that underground hydrothermal reservoirs exist on the Red Planet. Once such reservoirs are found, they can be used to supply the settlers with abundant supplies of both water and geothermal power. As more people steadily arrive and stay longer before they leave, the population of the town will increase. In the course of things, children will be born and families raised on Mars, the first true colonists of a new branch of human civilization.

We don't need any fundamentally new or even cheaper forms of interplanetary transportation to send the first teams of human explorers to Mars. However, meeting the logistical demands of a Mars base will create a market that will bring into being low-cost, commercially developed systems for interplanetary transport. Combined with the base's own activities in developing the means to use Martian resources to allow humans to be self-sufficient on the Red Planet, such transportation systems will make it possible for the actual colonization and economic development of Mars to begin.

(See plate 8.)

While the initial exploration and base-building activities on Mars can be supported by government or corporate largesse, a true colony must eventually become economically self-supporting. Mars has a tremendous advantage compared to the moon and asteroids in this respect, because unlike these other destinations, the Red Planet contains all the necessary elements to support both life and technological civilization, making self-sufficiency possible in food and all basic, bulk, and simple manufactured goods.

That said, Mars is unlikely to become autarchic for a very long time, and even if it could, doing so would not be advantageous. Just as nations on Earth need to trade with each other to prosper, so the planetary civilizations of the future will also need to engage in trade. In short, regardless of how self-reliant they may become, the Martians will always need, and certainly always want, cash. Where will they get it?

A variety of ideas have been advanced for potential cash exports from Mars. For example, Mars might serve as a source of food and other useful goods for asteroid-mining outposts that themselves export precious metals to Earth. Or, since the water on Mars has six times the deuterium concentration as Earth's, that potentially very valuable fusion-power fuel could be exported to the home planet once fusion power becomes a reality. Such deuterium could be extracted as a profitable by-product of routine Martian life support and fuel-making operations, which require electrolyzing water into hydrogen and oxygen as one step in their processes. Or maybe precious metals will be found on Mars that, with a fully reusable interplanetary transportation system, might be profitable to mine and export to Earth. Yet another option is simply selling land. Just as during the settlement of the American West, undeveloped land can have speculative value based on its future prospects for development, with the highest prices obtained for parcels close to a settled base, transportation route, or actual or potential (for example, geothermal) power source, or otherwise rich in an identifiable resource.

While such possibilities exist, in my view the most likely export that Mars will be able to send to Earth will be patents. The Mars colonists will be a group of technologically adept people in a frontier environment where they will be free to innovate—indeed, *forced* to innovate—to meet their needs, making the Mars colony a pressure cooker for invention. For example, the Martians will need to grow all their food in greenhouses, strongly accentuating the need to maximize the output of every square meter of crop-growing area. They thus will have a powerful incentive to engage in genetic engineering to produce ultraproductive crops and will have little patience for those who would restrict such inventive activity with fearmongering or red tape.

Similarly, there will be nothing in shorter supply in a Mars colony than human labor time, and so just as the labor shortage in nineteenth-century America led Yankee ingenuity to a series of labor-saving inventions, the labor shortage on Mars will serve as an imperative driving Martian ingenuity in such areas as robotics and artificial intelligence. Recycling technology to recapture valuable materials that would otherwise be lost as wastes will also be strongly advanced. Such inventions, created to meet the needs of the Martians, will prove invaluable on Earth, and the relevant patents, licensed on Earth, could produce an unending stream of income for the Red Planet. Indeed, if the settlement of Mars is to be contemplated as a private venture, the creation of such an inventor's colony—a Martian Menlo Park—could conceivably provide the basis for a fundable business plan.

To those who ask what are the "natural resources" on Mars that might make it attractive for settlement, I answer that there are none, but that is because there are no such thing as natural resources anywhere. There are only natural raw materials. Land on Earth was not a resource until human beings invented agriculture, and the extent and value of that resource has been multiplied many times as agricultural technology has advanced. Oil was not a resource until we invented oil drilling and refining as well as technologies that could use the product. Uranium and thorium were not resources until we invented nuclear fission. Deuterium is not a resource yet but will become an enormous one once we develop fusion power, an invention that future Martians, having limited alternatives, may well be the ones to bring about. Mars has no resources today but will have unlimited resources once there are resourceful people there to create them.

Martian civilization will become rich because its people will be smart. It will benefit the Earth not only as a fountain of invention but as an example of what human beings can do when they rise above their animal instincts and invoke their creative powers. It will show all that infinite possibilities exist—not to be taken from others, but to be made.

In addition to inventions, Mars will eventually also have material products for export, to the Earth, the moon, or outposts in the asteroid belt and beyond. Martian colonists will be able to use reusable rocket hoppers using locally produced propellants to lift such resources from the Martian surface to Mars's moon Phobos, where an electromagnetic catapult can be emplaced capable of firing the cargo off for interplanetary export. Larger or more complex cargoes could be shipped out from Phobos at low cost using solar-sail-powered robotic spacecraft.

Alternatively, a skyhook tether could be hung 5,800 kilometers down from Phobos, which orbits at a speed of 2.14 km/s 6,000 kilometers above the Martian surface, which itself is 3,400 kilometers from the planet's center. This being the case, the bottom of the tether would be traveling at a speed of just 0.82 km/s above the Martian equator, which itself is turning in the same direction at a speed of 0.24 km/s. As a result, a rocket vehicle taking off from the equator would be able to reach the tether with a ΔV of just 0.58 km/s, catch it, and then be hauled up to Phobos by a cable car. If there were another cable extending outward from Phobos, the vehicle could be *lowered outward* (sic—along this cable, effective gravity would point away from Mars, since centrifugal force would be greater than gravity) to reach escape velocity at 3,726 kilometers beyond Phobos and still greater speeds at distances beyond. Assuming a tether tensile strength equal to Kevlar (2,800 megapascals), the tether would have a mass less than ten times the payload it could lift. By using such a system, cargo could be sent back to Earth or whipped out to the asteroid belt or beyond with a tenth of the rocket ΔV required to do so directly. This would make it feasible for Mars colonists to transport goods cheaply to Earth; the moon; the asteroid belt; and the moons of Jupiter, Saturn, Uranus, and Neptune. High-technology goods needed to support asteroid mining may have to come from Earth for some time. But since food, clothing, and other necessities can be produced on Mars with much greater ease than would be possible anywhere further out, Mars could become the central base and port of call for exploration and commerce heading out to the asteroid belt, the outer solar system, and beyond.

TERRAFORMING MARS

Currently available scientific evidence shows that Mars was once a warm and wet planet, a place friendly to life. Atmospheric resources in the form of vast reserves of carbon dioxide and water adsorbed or frozen into the soil still exist. Once sufficient human industrial potential is developed on Mars, human colonists might begin to employ it to return the planet to the warm-wet climate of its distant past. By producing fluorocarbon (CF) super greenhouse gases on Mars at a rate similar to what we are now doing on Earth, and willfully dumping these climate-altering substances into the atmosphere, Mars colonists could, over a period of several decades, warm the planet by as much as 10°C. This warming would have the effect of causing massive quantities of carbon dioxide to outgas from the soil. Since CO_2 is a greenhouse gas, this would raise the temperature of the planet still more. As the temperature rises, the vapor pressure of water in the Martian atmosphere would also rise, and since water vapor is also a very strong greenhouse gas, this would raise the temperature of the planet still more, forcing even more CO_2 out of the soil, and so on. As a result of these positive feedback mechanisms, a runaway greenhouse effect could be created on a planetwide scale on Mars, with the net result being to raise the average temperature on the planet by more than 50°C within half a century.[10] At the same time, the atmospheric pressure would rise from its current level of about 1 percent that of Earth to about 35 percent (i.e., five pounds of pressure per square inch).[11] An air pressure of 5 psi may not sound like much, but it's what we used on the Skylab space station in the early 1970s. Provided that the atmosphere is enriched to be 60 percent oxygen and 40 percent nitrogen (instead of the 20 percent oxygen/80 percent nitrogen that prevails on Earth), such gas is perfectly breathable. So, while humans could not breath the 5-psi CO_2 atmosphere that would prevail on such an altered Mars, they no longer would need to wear space suits—simple breathing gear providing oxygen-enriched gas would suffice. Available habitation volume would expand dramatically as well, since the availability of 5-psi external pressure would allow very large inflatable domes featuring an internal breathable 5-psi atmosphere (3 psi oxygen/2 psi nitrogen) to be readily erected. Moreover, as the outside environment would be warm enough for liquid water, the Martian permafrost would start to melt, and plants could propagate across first the tropical, then the temperate regions of Mars. Over a period of a thousand years or so—or potentially much faster using genetically engineered organisms—such plants could put enough oxygen in Mars's atmosphere to make it breathable by humans and higher animals. Eventually, the day would come when the breathing gear and city domes would no longer be necessary.

This feat, the "terraforming" of Mars from its current lifeless or near-lifeless state to a living, breathing world supporting multitudes of diverse and novel life-forms and ecologies, will be one of the greatest and noblest enterprises of the human spirit. No one will be able to contemplate it and not feel prouder to be human.

Life in the initial Mars settlements will be harder than life on Earth for most people, but life in the first North American colonies was much harder than life in Europe as well. People will go to Mars for many of the same reasons they went to colonial America: because they want to make a mark, or to make a new start, or because they are members of groups who are persecuted on Earth, or because they are members of groups who want to create a society according to their own principles. Many kinds of people will go, with many kinds of skills, but all who go will be people willing to take a chance to do something important with their lives. Out of such people are great projects made and great causes won. Aided by ever-advancing technology, such people can transform a planet and bring a dead world to life.

FOCUS SECTION: PLANETARY PROTECTION?

A persistent objection to human Mars exploration from some quarters is the need for "planetary protection."

The story goes like this: No Earth organism has ever been exposed to Martian organisms, and therefore we would have no resistance to diseases caused by Martian pathogens. Until we can be assured that Mars is free of harmful diseases, we cannot risk infecting the crew with such a peril that could easily kill them or, if it didn't, return to Earth with the crew to destroy not only the human race but the entire terrestrial biosphere.

The kindest thing that can be said about the above argument is that it is just plain nuts. In the first place, if there are or ever were organisms on or near the Martian surface, then the Earth has already been, and continues to be, exposed to them. The reason for this is that over the past billions of years, trillions of tons of Martian surface material have been blasted off the surface of the Red Planet by meteor strikes, and a considerable amount of this material has traveled through space to land on Earth. We know this for a fact because scientists have collected nearly a hundred kilograms of a certain kind of meteorite called "SNC meteorites" and compared the isotopic ratios of their elements to those measured on the Martian surface by the Viking landers. The combinations of these ratios (things like the ratio of nitrogen-15 to nitrogen-14), as well as the fact that the gas trapped in the rock matches the Martian atmosphere, represent an irrefutable fingerprint proving that these materials originated on Mars. Despite the fact that in general each SNC meteorite must wander through space for millions of years before arrival at Earth, it is the opinion of experts that neither this extended period traveling through hard vacuum nor the trauma associated with either initial ejection from Mars or reentry at Earth would have been sufficient to sterilize these objects, if they had originally contained bacterial spores. Indeed, chemical investigations of the famous SNC meteorite ALH84001 have shown that portions of it never rose above 40°C at any time during its entire interplanetary journey, and therefore, if there had been bacteria within it when it departed Mars, they easily would have survived the trip. Furthermore, on the basis of the amount we have found, it has been estimated that these Martian rocks continue to rain down upon the Earth at a rate of about five hundred kilograms per year. So, if you're scared of Martian germs, your best bet is to leave Earth fast, because when it comes to Martian biological warfare projectiles, this planet is smack in the middle of Torpedo Alley. But don't panic—they're not so dangerous. In fact, to date, the only known casualty of the Martian barrage is a dog who was killed by one of the falling rocks in Nakhla, Egypt, in 1911. Statistically, the hazard presented to pedestrians by furniture being thrown out onto the street from upper-story windows is a far greater threat.

So the idea of quarantining rocks brought back from Mars by astronauts or a robotic sample return mission makes about as much sense as having the border patrol inspect incoming cars to make sure they are not transporting Canada geese into the country. They are flying in all the time on their own.

The fact of the matter, however, is that life almost certainly does not exist on the Martian surface. There is no (and cannot be) liquid water there—the average surface temperature and atmospheric pressure will not allow it. Moreover, the planet is covered with oxidizing dust and bathed in ultraviolet radiation to boot. Both of these last two features—peroxides and ultraviolet light—are commonly used on Earth as methods of sterilization. No, if there is life on Mars now, it almost surely must be ensconced in underground water.

But couldn't such life, if somehow unearthed by astronauts, be harmful? Absolutely not. Why? Because disease organisms are specifically keyed to their hosts. Like any other organism, they are expressly adapted to life in a particular environment. In the case of human disease organisms, this environment is the interior of the human body or that of a closely related species, such as another mammal. For almost four billion years, the pathogens that afflict humans today waged a continuous biological arms race with the defenses developed by our ancestors. An organism that has not evolved to breach our defenses and survive in the microcosmic free-fire zone that constitutes our interiors will have no chance of successfully attacking us. This is why humans do not catch Dutch elm disease and trees do not catch colds. Now, any indigenous Martian host organism would be far more distantly related to humans than elm trees are. In fact, there is no evidence for the existence of, and every reason to believe the impossibility of, macroscopic Martian fauna and flora. In other words, without indigenous hosts, the existence of Martian pathogens is impossible, and if there were hosts, the huge differences between them and terrestrial species would make the idea of common diseases an absurdity. Equally absurd is the idea of independent Martian microbes coming to Earth and competing with terrestrial microorganisms in the open environment. Microorganisms are adapted to specific environments. The notion of Martian organisms outcompeting terrestrial species on their home ground (or terrestrial species overwhelming Martian microbes on Mars) is as silly as the idea that sharks transported to the plains of Africa would replace lions as the local ecosystem's leading predator.

If I appear to be spending excessive time on refuting this idea, it's partly as a result of a NASA planning meeting for the planned (robotic) Mars sample return mission during which someone seriously proposed that, to allay alleged public concerns, any sample acquired on Mars be sterilized by intense heat before being brought to Earth. While an extremely unlikely find, the greatest possible treasure a Mars sample return mission could provide would be a sample of Martian life. Yet certain of those attending the meeting would destroy it preemptively (and a great deal of valuable mineralogical information in the sample as well). The proposal was so grotesque that I countered by asking the assembled scientists, "If you should find a viable dinosaur egg, would you cook it?" The question is not entirely out of line; after all, dinosaurs are our comparatively close relatives, and they did have diseases. In fact, every time you turn over a shovelful of dirt, you are returning a sample of the Earth's disease-infested past to menace the current biosphere. Nevertheless, neither paleontologists nor gardeners generally wear decontamination gear.

Just as the discovery of a viable dinosaur egg would represent a biological treasure trove but no menace, so a sample of live Martian organisms would be a find beyond price but certainly constitute no threat. In fact, by examining Martian life, we would have a chance to differentiate between those features of life that are idiosyncratic to terrestrial life and those that are generic to life itself. We could thus learn something fundamental about the very nature of life. Such basic knowledge could provide the basis for astonishing advances in genetic engineering, agriculture, and medicine. No one will ever die of a Martian disease, but it might be that thousands of people are dying today of terrestrial ailments whose cure would be apparent if only we had a sample of Martian life in our hands.

The alternative form of the planetary protection argument is the purported need *not* to go to Mars in order to be able to prove that any microbes we find there are in fact indigenous, and not simply bugs transported with the astronauts themselves. This is also silly.

There are two possibilities. Either microbes found on Mars are of the types we are familiar with on Earth, which all use the same RNA-DNA alphabet for transmitting genetic information, or they are different. If they use a different genetic alphabet, then we didn't bring them. If they use the same genetic alphabet but are of types never before observed on Earth despite four centuries of searching by microbiologists, then also, we clearly didn't bring them.

If we find microbes in places where we hadn't been at all—for example, underground reservoirs—then, even if they were similar to Earth bugs, it would be clear that we hadn't brought them.

But what if all we find are microbes in places where we have been, and they are the same types that we see on Earth—for example, E. coli? How could we know then?

It's simple. If microbes were on Mars before we arrived, they would, by definition, have been there in the past and left fossils or other biomarkers to prove it. Unless one believes, as, for example, the "creationists" do, that fossils are frauds planted in the ground by a god to test our faith in the Bible or other ancient texts, rather than proof of the existence of a geologic past, such evidence is conclusive.

On the other hand, if all you do is find familiar bugs in places where you've been, with no evidence that they existed on Mars prior to your arrival, well, then you will just need to keep searching before you get your Nobel Prize.

So the argument that we owe it to science not to go to Mars is simply wrong.

If we go, we can know. If we don't go, we won't know.

Seek and you shall find. The truth will set you free.

Chapter 5

ASTEROIDS FOR FUN AND PROFIT

I would rather believe two Yankee professors would lie,
than that stones have fallen from the heavens.
                                    —Thomas Jefferson

It is a general law of nature that for virtually every type of thing, there are more little ones than big ones. There are more dandelions than towering redwoods, more minnows than sharks, more mice than elephants, more sand grains than pebbles, and more pebbles than boulders.

Thus it should have come as no surprise that there are far more tiny planets than grand ones, and many more interplanetary free-floating rocks than miniplanets—some of which might well collide with the Earth and "fall from the heavens" from time to time. Yet, surprise it was when on January 1, 1801, Sicilian astronomer Giuseppe Piazzi opened the new century by discovering a miniature planet, which he named Ceres, orbiting the sun between Mars and Jupiter.

While displeasing to philosophers, who thought that the seven known planets (Uranus had been discovered by William Herschel in 1781) comprised precisely the right number, the discovery of Ceres was very gratifying to astronomers, particularly as its orbital distance from the sun, 2.7 astronomical units, was very close to the 2.8 predicted for a missing planet by the German astronomers Titius and Bode based upon other planetary orbit distance ratios back in the 1770s. (One astronomical unit, or AU, equals 150 million kilometers, the distance at which the Earth orbits the sun.) Ceres proved to be a very small planet, only nine hundred kilometers in diameter (about one-quarter that of the Earth's moon), but this disappointment was compensated for when, over the next few years, astronomers discovered planetoids Pallas, Juno, and Vesta in similar orbits close by. This suggested to the

German astronomer Heinrich Olbers that the four objects were fragments of a properly sized planet that had broken up. More fragments could therefore be expected, and as the nineteenth century wore on, and telescopes got better, this hypothesis was substantiated as dozens and then hundreds of additional "asteroids" were discovered. By 1890, more than three hundred were known, all orbiting the sun in a belt between 2 and 3 AU, nicely bracketed near the geometric mean of 2.8 AU between the orbit of Mars (1.52 AU) and that of Jupiter (5.2 AU).

But then, in 1898, an asteroid was discovered whose maximum distance, or aphelion, from the sun is 1.78 AU, and whose minimum distance, or perihelion, is only 1.14 AU. It thus *crosses the orbit of Mars* and sometimes swings within twenty million kilometers of the Earth. This wandering across orbits was considered truly errant behavior, and so to distinguish it from the well-behaved female deities orbiting the Main Belt so nicely, the new ten-kilometer-class asteroid was given a male name, Eros.

Figure 5.1. The near-Earth asteroid Eros, photographed by NASA's NEAR spacecraft in 2000. *Image courtesy of NASA.*

Soon other male, planet-crossing asteroids were discovered, and some of these crossed not only Mars's orbit but that of Earth as well. In 1932, the asteroid Apollo crossed the Earth's orbit and passed within ten million kilometers of our world. In 1936, Adonis passed us at a distance of two million kilometers. On March 23, 1989, the "small" (eight hundred meters, or 120 million kilotons of impact energy) and therefore unnamed asteroid 1989 FC swept by at a distance of 720,000 kilometers, passing through a point in space that the Earth had occupied less than six hours before.

As mentioned above, about two hundred Earth-crossing, "male," or "near-Earth" asteroids larger than one kilometer are known today, and it is estimated that there are at least two thousand of them out there. More than seven thousand "female" asteroids are known to exist in the Main Belt between Mars and Jupiter, including every asteroid larger than ten kilometers, hundreds larger than one hundred kilometers, and one as large as nine hundred kilometers. Because they orbit farther from both the sun and the Earth, the smaller female Main Belt asteroids are much harder to see than their Earth-crossing counterparts. It's estimated that there are at least two million Main Belt asteroids larger than a kilometer.[1] The girls thus outnumber the boys by about a thousand to one, and utterly dwarf them as well. It's good the lasses are well-behaved. If they acted like their brothers, all life on Earth would have been exterminated a thousand times over.

Of course, there is a reason for the overwhelming dominance of female asteroids: Male asteroids don't live long. They kill themselves by crashing into us.

Such events have been very inconvenient to Earth's previous inhabitants, many of whom have been wiped out as a result. In a subsequent chapter, we will discuss what we can do to prevent them.

The total mass of all the asteroids adds up to only about 4 percent that of the moon, or 0.05 percent that of the Earth. Yet because of their numbers, the total surface area of all the asteroids—*in their current form*—is about the same as the moon, or the continent of Africa. That's sizable, but it considerably understates the possibilities.

The Earth may have a radius of nearly 6,400 kilometers, but at most, only the top 6 kilometers, or 0.1 percent, is within reach. In contrast, the material contents of most asteroids are fully accessible. If reconfigured for human habitation, the mass of a 1-kilometer asteroid could afford living space equal to that of a major city. For example, if rebuilt as a structure resembling one of Gerard O'Neill's space colonies, a single 1-kilometer-radius asteroid would provide enough mass for a rotating cylinder 1 kilometer in diameter, 1 meter thick, and 1,250 kilometers long, or about 4,000 square kilometers. For comparison, London (population 8.2 million) has an area of 1,623 square kilometers. Approached this way, the total potentially habitable territory of the Main Belt would be about a hundred times that of the Earth, dispersed amid the volume of about seventeen cubic AU, or 230,000 billion trillion (sic) cubic kilometers. It's a vast ocean, filled with millions of worlds, which could someday be homes to millions of new city-states, nations, and civilizations.

ASTEROID EXPLORATION

In one sense, we already know more about the asteroids than nearly any other extraterrestrial body, except perhaps the moon, because we have hundreds of thousands of samples. These are the meteorites, fragments of asteroids that have—despite Thomas Jefferson's disbelief—fallen to Earth and are available for collection. The meteorites show a variety of asteroid compositions, ranging from nearly pure metal, to stone, to carbonaceous materials. Because they survive atmospheric entry best, and because they are easiest to distinguish from terrestrial rocks, iron-metal meteorites enjoy preferential representation in terrestrial meteorite collections. However, by comparing the spectral characteristics of meteorites with those reflected from asteroids in space, astronomers have been able to classify the asteroids according to their composition. The principal types are listed in the table below.

TABLE 5.1. PRINCIPAL TYPES OF ASTEROIDS

Type	Composition	Primary concentration
M	Metal	Inner Main Belt, near Mars
E	Silicate rocks	Inner Main Belt, 1.9 AU
S	Stony-iron	Central Main Belt, 2.4 AU
C	Carbonaceous	Outer Main Belt, 3.3 AU
P	Carbonaceous/volatile	Outermost Main Belt, 4 AU
D	Frozen volatiles	Beyond Jupiter

No type of asteroid is concentrated in the near-Earth objects, because the NEOs represent only a tiny minority of the asteroids overall. Also, because they are small, it is hard to get a composition-determining reflection from most of the known NEOs. However, of those that have been surveyed by astronomer Lucy-Ann McFadden, about 80 percent were found to be S asteroids and 20 percent were C asteroids. Types M and E were probably not observed simply due to the small size of the assessed sample: these are generally rarer than S and C types. Types P and D cannot exist in near-Earth space for long without evaporating. If one came our way, we would observe it as a comet.

With the exception of radar imaging of a few close-passing NEOs (such as Toutatis, which was imaged by the Jet Propulsion Lab's Steve Ostro using the Goldstone Deep Space Communications Complex as radar during the asteroid's 3.5-million-kilometer pass by our planet in December 1992), asteroids are too small or too far away to be photographed by Earth-based telescopes. Our first good look at asteroids, therefore, had to wait for images returned by interplanetary spacecraft. The first of these was produced by the Galileo spacecraft when it flew by the Main Belt asteroid Gaspra on its way to Jupiter in October 1991. Galileo's magnetometer also measured a surprisingly strong magnetic field around the asteroid, indicating the presence of a large quantity of metallic iron. The images returned show Gaspra to be a potato-shaped object, nineteen kilometers long by eleven in diameter. Based on counting the number of craters on its surface, mission scientists judged that Gaspra is only about four hundred million years old, which is a bit of a mystery since the rest of the solar system is ten times that age. Perhaps Gaspra is a fragment cast off by a catastrophic collision between two larger bodies four hundred million years ago.

In August 1993, Galileo, pushing further out through the Main Belt, was able to get a look at another asteroid, fifty-one-kilometer-long Ida. Surprisingly, Ida was found to have a moon of her own, a one-kilometer-sized object, which mission scientists named Dactyl. Since the rate of Dactyl's orbit about Ida is dependent on Ida's mass, and since Ida's size is known, the mission team was able to use measurements of Dactyl's orbit to calculate that the density of Ida is about 2.5 times that of water, consistent with a carbonaceous composition.[2]

But these Galileo flybys were just quick snapshots. Much more can be learned by having a dedicated spacecraft rendezvous with an asteroid and hanging around to take close-up photographs and detailed sets of measurements. This was done by the Near Earth Asteroid Rendezvous (NEAR) mission, designed by the Johns Hopkins Applied Physics Lab, which reached and orbited Eros, the first discovered and largest of the NEOs, in January 1999. NEAR orbited Eros for more than a year, gradually lowering its orbit, and then actually *landed* on it in 2001, returning spectacular photographs in the process.[3]

This extraordinary achievement was then exceeded by the Japanese Hayabusa mission, which not only landed on the asteroid Itokawa but took off again to return dust samples to Earth in 2010.[4] The European Rosetta spacecraft flew by asteroids Steins in 2008 and Lutetia in 2010 on its way to orbit the comet Churyumov-Gerasimenko in 2014.[5] Another notable mission was the JPL Dawn spacecraft, a tour de force that, after orbiting the large asteroid Vesta from 2011 to 2012, used very-high-exhaust-velocity electric propulsion to depart and then go into orbit around the dwarf planet Ceres in 2015.[6]

The most recent asteroid missions are the Japanese Hayabusa2 and NASA's OSIRIS-REx, which launched in December 2014 and September 2016 to explore the carbonaceous asteroids Ryugu and Bennu, respectively. Hayabusa2 arrived at Ryugu in July 2018 and subsequently deployed the small European Mobile Asteroid Surface Scout (MASCOT) lander to collect samples.[7] In December 2019, it will depart and hopefully return its samples to Earth in December 2020. OSIRIS-REx reached Bennu in August 2018 and is now in orbit mapping the asteroid.[8] If all goes well, it will land on Bennu in July 2020 and use a robotic arm to secure at least sixty grams of samples, which it will return to Earth via a reentry capsule in September 2023.

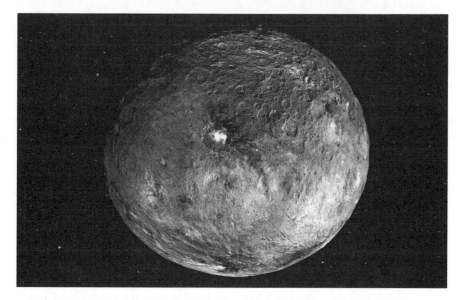

Figure 5.2. The Main Belt asteroid Ceres photographed by the NASA/JPL Dawn spacecraft in 2015. The bright spots are believed to be water ice. *Image courtesy of NASA.*

GAIASHIELD: A HUMAN ASTEROID MISSION

However, as impressive as these robotic probes may be, they have only scratched the surface. Given the importance of gaining a detailed knowledge of NEOs to the future security of humanity, the costs associated with human exploration are more than justified. If we were to launch the Mars Direct program described in chapter 4, the same launch vehicles and hab modules developed for Red Planet exploration could also be used to perform rendezvous and return missions to near-Earth asteroids. Indeed, since the rocket propulsion requirements required to leave low Earth orbit (LEO) for a one-way trajectory and landing on Mars (a ΔV of about 4.7 km/s) are nearly identical to those for a round-trip from LEO to many NEOs, hardware designed for Mars Direct–type missions enjoys natural commonality with that needed for NEO exploration. But because asteroids have no atmosphere and little gravity, eliminating the need for reentry and landing systems, and because asteroids are small, eliminating the need for ground vehicles and split field-exploration and base crews, a minimal, piloted asteroid exploration mission can be launched with a significantly smaller and more limited set of hardware than that needed for Mars exploration. Such a mission could be flown within four years, using launch vehicles and technology available today. If there is no manned Mars exploration program, it may help produce one, because in the course of flying to an asteroid and back, the astronauts will destroy the putative barriers to long-distance spaceflight of cosmic rays, subnormal gravity, and human factors, which are used as excuses for lack of initiative by apologists for current go-nowhere space policies.

I call my asteroid mission plan "Gaiashield," because it would be an important first step in giving humanity the knowledge and spacefaring capability we need to protect the Earth's biosphere from another mass extinction.

The Gaiashield mission could employ a simple cylindrical habitation module, five meters in diameter and twenty meters long, somewhat similar to those used on the International Space Station. Alternatively, an inflatable hab module of the Bigelow type could be used. This module could be launched in one piece, together with a space-storable methane/oxygen chemical propulsion stage, by a Falcon Heavy, New Glenn, or SLS launch vehicle. The module would be equipped with a set of photovoltaic panels, which would deploy outward from it like wings, somewhat in the manner of the United States' 1970s-era space station, Skylab. After it reached orbit, a Falcon 9 would be used to deliver the crew to it in a Dragon capsule, which would remain mated to the hab for use as a reentry and landing vehicle at the end of the mission. Once the crew has determined that all is shipshape for Earth departure, one more Falcon Heavy–class launcher would send a fifty-ton high-energy hydrogen/oxygen propulsion stage to mate with the stack and boost it into a highly elliptical, near-escape-velocity orbit around the Earth. (The H_2/O_2 stage would be sent last because its propellant is not readily space storable. If a space-storable stage were used instead, the schedule would be more flexible, but the payload would need to be reduced.) Using their methane/oxygen propulsion system, the crew would then boost themselves beyond Earth escape to a trans-asteroid trajectory. The H_2/O_2 stage could then be directed to gently aerobrake itself back to LEO for potential reuse by executing a series of passes through the Earth's upper ionosphere.

Once on the trans-asteroid trajectory, small reaction control thrusters would fire, causing the spacecraft to spin in the same plane in which the solar panels are located, with the spin axis and the solar panels pointing at the sun. The length of the spin arm between the spacecraft center of gravity and the decks at the far end of the module would be about ten meters. As a result, spinning at 4 RPM, lunar-equivalent gravity could be generated at the "lowest" decks. Spinning at 6 RPM, Mars-level gravity could be created. While NASA officials wishing to justify space station research programs on the human health effects of long-duration zero-gravity exposure frequently affect deep concern over the possible disorientation caused by Coriolis forces and other concomitants of artificial gravity systems, experiments done in the 1960s show that humans can adapt and operate well in vehicles rotating as rapidly as 6 RPM. Many current artificial gravity researchers, such as Professor Larry Young of MIT, believe that rotation rates as high as 10 RPM are viable. With artificial gravity, the Gaiashield crew would be protected against the severe negative health impacts that have afflicted cosmonauts and others who failed to implement strenuous exercise programs when flying for long periods in zero gravity.

The ship would take about six months to reach the asteroid, as despite its close distance to the Earth, an elliptical trajectory from one side of Earth's orbit to the other would probably be needed to get there. Upon interception, the crew would despin the ship and use most of the propellant in the propulsion stage to effect a ΔV of perhaps 0.5 km/s to establish the spacecraft in orbit a few kilometers from the asteroid, where it will remain for one year. The crew would then proceed to explore the asteroid in detail, using backpack gas thrusters similar to the space shuttle's Manned Maneuvering Unit to fly to it from the ship and hop about the body at will. A small portable drilling rig would be used to take repeated deep samples from all over the body.

At the end of a year of intensive exploration, the propulsion stage would fire its last allotment of propellant to give the ship the required 0.5 km/s ΔV needed to send it on trans-Earth injection. After a voyage of another six months, the ship would approach Earth, and the crew would bail out in the reentry capsule and be picked up by a boat, much as the Apollo astronauts did a half century ago. Empty of crew, the ship itself would remain in a cycling orbit between the Earth and the asteroid, possibly to be refitted for further use when appropriate capabilities for doing so are developed.

The crew would thus spend two years in interplanetary space, about twice that required for a round-trip to Mars (which spends six months traveling each way and 1.5 years on the surface). During this trip, they would take about 100 rem of cosmic radiation, which represents about a 2 percent statistical risk of fatal cancer later in life for each member of the crew. (In contrast, an average smoker incurs a 20 percent risk.) This is small compared to other risks associated with piloted space missions, and there is no doubt that many astronauts would be more than willing to take it on.

Thus, using two launches of a Falcon Heavy, New Glenn, or SLS booster, plus one launch of a Falcon 9 with a Dragon, a Spartan two-person human mission to a near-Earth asteroid in interplanetary space could be readily accomplished.

However, once the SpaceX Starship is available, it could be used to fly a crew of twenty or more to a near-Earth asteroid in style. The mission would be quite similar to the one SpaceX has announced it will use to send a crew plus a group of eight artists in a flight around the moon circa 2023, except that because the mission will take two years instead of one week, a lot more supplies will need to be brought along. But with the Starship's 150-ton-to-orbit payload capability, this should not be a problem.

The mission plan would be quite simple. First a Starship with its supplies and crew would be launched to orbit. Then two more cargo Starships would be used to deliver three hundred tons of propellant to refuel the mission craft. Then it's off for the two-year round-trip voyage to the asteroid, with plenty of room and company aboard to enjoy the trip.

(See plate 9.)

Starship is not an artificial gravity spacecraft, but if two were flown, they could tether off each other nose to nose. If the tether were five hundred meters long, spinning the assembly at 2 RPM would create Earth-level gravity in each ship during cruise. Upon arrival at the asteroid, thrusters could be used to stop the spin, making exploration operations more convenient and enabling zero-gravity concerts and other recreations. Taking two ships in this way would also provide 100 percent mission backup if anything should go wrong. Alternatively, just one ship could be used, with artificial gravity enabled by tethering off a counterweight brought along for the purpose.

Whether done in a small current technology hab module or a luxurious Starship cruiser, Gaiashield would be a terrific asteroid science mission, but it would be more: it would be an *icebreaker* mission. Two things have kept NASA from sending human explorers to Mars. The first is the notion that such missions must be incredibly expensive. The second is fear of the risks involved. These two factors have fed off each other, with fear of long-duration space voyages making NASA place the Mars mission at the end of an impossibly expensive multidecade series of preparatory activities.

The debilitating effects of long-duration spaceflight are not caused by radiation. No astronaut or cosmonaut has ever received a radiation dose during flight large enough and prompt enough to create any visible effects. Rather, all the well-known ill effects are due to long-duration zero-gravity exposure and ensuing complications.

The Gaiashield mission demonstration of a piloted interplanetary spacecraft with artificial gravity would kill forever the dragons of cosmic ray threat and of zero-G space sickness that are barring us from the solar system. Furthermore, it would destroy the myth that interplanetary manned exploration need be impossibly costly. It would also directly accomplish most of the nonrecurring development that needs to be performed for a human Mars mission.

Before Copernicus, Ptolemaic astronomers believed that humanity was walled off from the heavens by a set of crystal spheres. In a way, those spheres are still there, made not of glass but of fear. The Gaiashield mission would smash them.

MINING THE ASTEROIDS

Most of the sensational attention to asteroids has centered primarily on their potential threat to humanity and the rest of the terrestrial biosphere. But just as fire, a deadly menace to animals and children who do not understand it, becomes in the hands of competent adults one of humanity's greatest boons, so the asteroids, which offer nothing but mass death for the pre-sentient biosphere or earthbound Type I humanity, hold the promise of vast riches for a Type II spacefaring civilization.

The asteroid belt is known to contain enormous supplies of very high-grade metal ore in a low-gravity environment that makes it comparatively easy to export to Earth. For example, in his book *Space Resources*, Professor John Lewis of the University of Arizona considers a single small type S asteroid just one kilometer in diameter—a run-of-the-mill asteroid. This body would have a mass of around two billion tons, of which 200 million tons would be iron, 30 million tons high-quality nickel, 1.5 million tons the strategic metal cobalt, and 7,500 tons a mixture of platinum group metals whose average value, at current prices, would be in the neighborhood of $40,000 per kilogram.[9] That adds up to $300 billion just for the platinum group stuff! There is little doubt about this—we have lots of samples of asteroids in the form of meteorites. As a rule, meteoritic iron contains 6–30 percent nickel, 0.5–1 percent cobalt, and platinum group metal concentrations at least ten times those of the best terrestrial ore. Furthermore, since the asteroids also contain a good deal of carbon and oxygen, all of these materials can be separated from the asteroid and from each other using variations of the carbon-monoxide-based chemistry needed for refining metals on Mars.

The economics of exporting asteroidal metals is worth examining.

First of all, it should be clear that for the mining economics to work, it is essential that refining take place. While the platinum group material may be worth $40,000 per kilogram, it only comprises 7,500/2,000,000,000 = 0.000375 percent of the bulk material, which therefore, based on its platinum content (alone), would only be worth $0.15/kg and not be worth shipping. At current prices ($0.70/kg for steel, $13/kg for nickel, $60/kg for cobalt, $40,000/kg for platinum group metals), it is clear that while asteroidal iron and nickel could have great value for use *in space*, we could produce something worth transporting for sale back on Earth only by refining the raw material down to, at most, the cobalt fraction plus the platinum group (which would create a combined 0.5 percent platinum group, 99.5 percent cobalt product worth about $260/kg).

**TABLE 5.2. VALUE OF COMPONENTS OF A TYPICAL
TWO-BILLION-TON TYPE S ASTEROID**

Component	Amount	Value/kg	Total value
Rock	1.8 billion tons	0	0
Iron	200 million tons	$0.70	$140 billion
Nickel	30 million tons	$13	$390 billion
Cobalt	1.5 million tons	$60	$90 billion
Platinum Group	7,500 tons	$40,000	$300 billion

But while we are at it, it would clearly be worth refining further, since if we could make it 10 percent platinum and 90 percent cobalt, the stuff would be worth $4,000/kg, allowing the miners to score a cool $400 million with each one-hundred-ton delivery to Earth. So, provided the right technology is put to work, the business case could well be there. But the easy money would not last forever.

If it were possible to obtain precious metals cheaply from the asteroid belt, such imports would flood the terrestrial market, driving down prices to well below current levels. While the profits of the first miners to get into the business could well be sky-high, such spectacular returns would swiftly draw competitors into the business. These would expand the quantity of imports until precious metal prices fell to the point where the rate of profit from asteroid mining would be no more than the average of comparable high-risk enterprises throughout the economy. At that point, prices would stabilize, turning asteroid mining into normal business, with further metal price drops caused mainly by technological improvements—which the existence of the business would itself drive.

So the net result of the asteroid platinum rush will most likely be giant fortunes for the daring first few, a steady income for many who follow, the creation of a host of new space technology enterprises, and a drastic lowering of the cost of many (now-) precious metals for everyone on Earth. The societal value of the last point cannot be understated. Platinum group metals are key ingredients for many promising new technologies—including fuel cells that could enable much better types of pollution-free electric cars than are possible using batteries. For example, a fuel cell car running on methanol (which can be produced from natural gas, coal, biomass, trash, or even CO_2 and water using nuclear or reusable energy) could be refueled as swiftly as a gasoline-powered car today, avoiding the long recharging time and high battery cost that is a major deterrent to the broad acceptance of electric vehicles. Cheap asteroidal platinum would make such vehicles extremely attractive, freeing terrestrial civilization from dependence on oil and eliminating automobile and many other types of urban air pollution, along with all the health problems they cause, everywhere.

But the terrestrial precious metal market is just the beginning. For every kilogram of platinum shipped back to Earth, thirty tons of excellent nickel-alloy steel would become available to build industries, spacecraft, habitats, and even free flying cities for multitudes of new branches of human civilization that will develop in space.

CLAIMING ASTEROIDS

The commercial potential of asteroid mining is so enormous that several start-ups have already been formed with the goal of pursuing the opportunity for profit. Among the leaders are Planetary Resources, begun by Peter Diamandis, Eric Anderson, and Chris Lewicki with backing by several Google magnates and other high rollers, and Deep Space Industries, founded by veteran space entrepreneurs David Gump and Rick Tumlinson, with noted asteroid expert Professor John Lewis serving as chief scientist. However, despite such strong founding groups, the hopes of these companies to realize their bold plans appear quite problematic, as there is currently no technically feasible way to mine the precious metals that exist in the asteroid belt and return them to Earth for sale. As a result, these firms are being forced to pursue more conventional aerospace technology applications in order to stay in business.

Yet this picture could be changed radically were a law enacted that would create a basis for private property claims in space. The asteroids, the moon, Mars, and other extraterrestrial bodies collectively contain vast areas of unexplored and potentially resource-rich territory. They carry no commercial value today, but this can be remedied swiftly.

Consider the following: Enormous tracts of land were bought and sold in Kentucky for large sums of money a hundred years before settlers arrived despite the fact that, for purposes of development, trans-Appalachian America in the 1600s might as well have been Mars. What made it salable were two things: (1) at least a few people believed that it would be exploitable someday; and (2) a juridical arrangement existed in the form of British Crown land patents, which allowed trans-Appalachian land to be privately owned.

Thus, if a mechanism were put in place that could enforce private property rights in space, mining claims probably could be bought and sold now. Such a mechanism would not need to employ enforcers (e.g., space police) patrolling the asteroid belt; the patent or property registry of a sufficiently powerful nation, such as the United States, would be entirely adequate.

For example, if the United States chose to grant a mining patent to a private group that surveyed an asteroid (or any other piece of extraterrestrial real estate) to some specified degree of fidelity, such claims would be tradable today on the basis of their future speculative worth, and probably could be used to privately finance robotic mining survey probes in the near future. Furthermore, such claims would be enforceable internationally and throughout the solar system simply by having the US Customs and Border Protection penalize with a punitive tariff any US import made anywhere, directly or indirectly, with material that was extracted in defiance of the claim.

This sort of mechanism would not imply American sovereignty over the solar system, any more than the current US patent and copyright offices' coining of ideas into intellectual property implies US government sovereignty over the universe of ideas. But, just as in the case of intellectual property, some government's agreement is needed to turn worthless terrain into real estate property value. The US Patent Office benefits inventors of every country by providing them a means of turning their creative efforts into negotiable property. In the same way, a US office for granting space mining patents would benefit all would-be planetary explorers, regardless of their nationality.

Once such a mechanism is in place, however, the undeveloped resources of space could become a tremendous source of capital to finance their exploration. Furthermore, when in private hands, the duly recognized claims would provide an incentive for their owners to further the development of technology that would enable their exploitation. As the capability for both exploration and development of space resources thus advanced, the value of both existing property claims and those obtainable in the future would increase, thereby expanding the financial resources available and accelerating space development even more.

The leaders and backers of companies like Planetary Resources and Deep Space Industries need to lobby hard to obtain legislation that would set up a legal regime for space property claims of this sort. The rest of us need to support its passage. Because if it does, massive new financial forces will be mobilized that will further the exploration and development of space.

With a stroke of a pen, a vibrant, privately funded space exploration effort could be brought into being, one that could use the daring and genius of the free market to rapidly bring the knowledge and the benefits of the vast untapped resources of the solar system to all humankind.

Lawmakers should take note and act accordingly.

THE SPACE TRIANGLE TRADE

While prospecting asteroids could become big business soon if the enabling legislation is passed, large-scale human activity in the Main Belt to exploit these claims will be difficult to support until we have a solid base on Mars. This is so because while water and carbonaceous material can readily be found among the asteroids (making them as a group far richer than the moon), it is not necessarily the case that such volatiles can be found on those asteroids that are most rich in exportable metals. Quite the contrary—the metal-rich type M asteroids are nearly volatile free. Moreover, while many of the Main Belt asteroids contain all the carbon, hydrogen, and oxygen needed to support agriculture, nitrogen is generally rare. Moreover, sunlight in the Main Belt is too dim to support agriculture, which means that plants would have to be grown by artificially generated light. This is a significant disadvantage for asteroid colonization, because plants are enormous consumers of light energy, and it is doubtful whether growing plants with electric lights to support any significant population is practical with current space power sources. Moreover, while collectively the asteroids may someday possess a significant mining workforce, until advanced robotic technology becomes available, it is unlikely that any one asteroid will have the sufficient personnel required to develop the division of labor necessary for true multifaceted industrial development.

Mining bases in the asteroid belt are a relatively near-term proposition. But farms, industries, and cities will need to wait until the widespread use of controlled fusion makes very large-scale employment of artificial power possible in the Main Belt. For the twenty-first century, most of the supplies needed to support the asteroid prospectors and miners will have to come from somewhere else.

As I showed in detail in my book *The Case for Mars*, even before a Phobos tether system is created, the ΔVs required to reach the asteroid Main Belt from Earth are more than double those required to access it from the Red Planet, leading to mass ratios at least seven times greater and mission gross liftoff masses *fifty times more*—and this is true whether chemical or electric propulsion systems are used.[10] And once the Phobos tether transportation system is put in place, the Martian payload delivery advantage increases more than an additional order of magnitude further.

The result that follows is simply this: anything that needs to be sent to the asteroid belt that *can* be produced on Mars *will* be produced on Mars.

The outline of mid-future interplanetary commerce in the inner solar system thus becomes clear. There will be a "triangle trade," with

Earth supplying high-technology manufactured goods to Mars; Mars supplying low-technology manufactured goods and food staples to the asteroid belt and possibly the moon as well; and the asteroids sending metals and perhaps the moon sending helium-3 to Earth. This triangle trade is directly analogous to the triangle trade of Britain, her North American colonies, and the West Indies during the colonial period. Britain would send manufactured goods to North America; the American colonies would send food staples and needed craft products to the West Indies; and the West Indies would send cash crops such as sugar to Britain. A similar triangle trade involving Britain, Australia, and the Spice Islands also supported British trade in the East Indies during the nineteenth century.

SETTLING THE ASTEROIDS

California, Nevada, and Colorado all entered American history as destinations for miners, drawn to those far-off locations by the promise of gold or silver. Ultimately, however, mining became an industry of secondary importance as settlers transformed mining outposts into towns, cities, and states. The same history may repeat itself in the Main Belt.

Collectively the asteroids have all the materials needed for life and civilization, but individually they do not. So clearly, what is required to enable settlement is technology readily enabling trade between them. Furthermore, as noted, asteroid settlement would need a lot of power, as the low solar fluence would make natural sunlight unattractive to support effective agriculture; a sevenfold concentrator would be needed to duplicate terrestrial illumination levels.

Fortunately, the Main Belt asteroids are rich in a material that can serve to meet both of these needs. That material is water.

One gallon of water on Earth contains enough deuterium that, if burned in a fusion reactor, it would produce as much energy as that released by combusting 350 gallons of gasoline. It is estimated that the deuterium content of asteroidal water, such as the ice the Dawn spacecraft imaged on the surface of Ceres, may contain twice as much. So, once we have deuterium fusion reactors, the power required to support asteroidal settlement could be met by a locally available fuel. Pure deuterium is not as attractive a fusion fuel as D-He3, as 40 percent of the reaction energy comes off as neutrons (which can induce radio-activity in the surrounding materials), but it will do the job until plentiful He3 from the outer solar system becomes available.

Fusion reactors could use water as a propellant, simply by heating it to high-temperature steam and exhausting it out a rocket nozzle to produce thrust. Such fusion thermal steam rockets could probably obtain an exhaust velocity of about 3.6 km/s (a specific impulse of about 350 seconds), giving them a performance similar to that of a kerosene/oxygen chemical rocket, but with a propellant that is easy to store and readily available throughout the Main Belt. Fission reactors could also be employed for this purpose, and in fact, such nuclear thermal rockets, using hydrogen propellant (thereby obtaining a spectacular exhaust velocity of 9 km/s), were developed and tested in the United States in the 1960s at sizes up to 250,000 pounds of thrust. But by employing fusion reactors to drive thermal rockets instead, the asteroid settlements could make themselves independent of Earth for fuel.

Alternatively, fusion reactors could be used to electrolyze water into hydrogen and oxygen. While far more difficult to produce and store than liquid water, these would allow spacecraft to travel among the asteroids using conventional lightweight chemical rockets, leaving the heavy reactors behind at propellant production bases.

With the aid of such systems, active trade among archipelagos of asteroidal settlements will become feasible, enabling them to collectively mobilize the division of labor and diverse material resources to give birth to a vast and vibrant space-based civilization.

NEW WORLDS FOR NEW SOCIETIES

The asteroids' multiplicity represents a disadvantage for societal development in the near term, but in the far term, it will be a great advantage. Mars, while huge, is after all one world. A multiple of social experiments will start there, but eventually these are likely to be resolved and fuse into a single, or at most a few, new branches of human civilization. But the technologies for resource utilization, labor saving, space transportation, and energy production developed for the colonization of Mars will open the way to the settlement of the asteroids, which will force both the technologies and the aptitudes that created them even further. This will make available thousands of potential new worlds, whose cultures and systems of law need never fuse.

Indeed, by far the greatest treasure that the asteroids offer humanity is not platinum but freedom. *There is nothing more valuable than freedom.*

Can space colonies truly be free? Some authors have argued that extraterrestrial liberty is unlikely, because the authorities in a space colony can always kill you by turning off your air.[11] But this has it backwards. Historically, the easiest people for a tyrant to oppress are nominally self-sufficient rural peasants, because none of them are individually essential. As the medieval saying went, "City air makes a man free." It is the interdependence and intercommunication of people in urban societies that empowers the individual. In a space colony, nearly everyone will be individually essential, and therefore powerful, and all will be capable of being dangerous to those in authority.

A society composed of very empowered citizens will need to treat its people right.

But freedom is more than the mere absence of tyranny. For nearly all human history, people have been required to live in realities and under rules that were completely defined before they were born. But the right to be a maker of your own world, rather than just an inhabitant of one, is a fundamental form of freedom. It only exists in a substantial way, however, in a society with an open frontier. One size will never fit all. There always will be people who have new ideas on how society should be organized, and if they are to be truly free, they need a place where they can go to give those ideas a try. The asteroid belt will provide home planets for thousands of such noble experiments.

Perhaps some will be republican, others anarchist. Some communalist, others capitalist. Some patriarchal, others matriarchal. Some aristocratic, others egalitarian. Some religious, others rationalist. Some Epicurean, others Stoic. Some hedonistic, others puritanical. Some traditional, others relentlessly innovative. Some may enthusiastically embrace transhumanism, augmented intelligence, and genetic engineering of children; others may reject them entirely. Those that afford people a chance to more fully realize their human potential will attract immigrants and grow. Those that do not will disappear. But there will no doubt be many diverse paths to success. As among the city-states of the ancient Greek islands, a bewildering myriad of diverse societies may flower and bloom—trading goods and ideas across a vast and endlessly creative cosmic cosmopolis.

The rest of humanity will watch and learn from their experiences. That which works will be repeated. So shall we continue to progress.

FOCUS SECTION: CHEMISTRY FOR SPACE SETTLERS

Just as the pioneers of old needed to know how to find the edible plants and methods of hunting the game available in their environments, so space settlers will need to know how to extract useful resources from their new worlds. The following is a brief compendium of some of the key techniques.

On the Moon

On the moon, oxygen can be produced from the mineral ilmenite, which is found in up to 10 percent concentrations in some lunar soils. The reaction is:

$$FeTiO_3 + H_2 \Rightarrow Fe + TiO_2 + H_2O \qquad (5.1)$$

The water produced is then electrolyzed to produce hydrogen, which is recycled back into the reactor, and oxygen, which, along with metallic iron, is the net useful product of the system. The feasibility of this system has been demonstrated by researchers working at Carbotek in Houston, Texas. If you don't want to go prospecting for ilmenite, you can try carbothermal reduction, a system pioneered by Sanders Rosenberg at Aerojet, which will work with a larger variety of lunar rocks, including the very common silicates.

$$MgSiO_4 + CH_4 \Rightarrow MgO + Si + CO + 2H_2O \qquad (5.2)$$

The water is then electrolyzed to produce oxygen, while the carbon monoxide and hydrogen from the electrolysis are combined to remake the methane in accord with:

$$CO + 3H_2 \Rightarrow CH_4 + H_2O \qquad (5.3)$$

Reactions 5.1 and 5.2 are very endothermic (i.e., they need energy input) and must be done at high temperatures (above 1,000°C). Reaction 5.3 is exothermic (i.e., it produces energy) and occurs rapidly at 400°C. The carbon and hydrogen reagents are extremely rare on the moon (except in permanently shadowed craters near the poles), so the systems must be designed for very efficient recycling.

On Mars

On Mars, the most accessible resource is the atmosphere, which can be used to make fuel, oxygen, and water in a variety of ways. The simplest technique is to bring some hydrogen from Earth and react it with the CO_2 that comprises 95 percent of the Martian air as follows.

$$CO_2 + 4H_2 \Rightarrow CH_4 + 2H_2O \qquad (5.4)$$

Reaction 5.4 is known as the Sabatier reaction and has been widely performed by the chemical industry on Earth in large-scale one-pass units since the 1890s. It is exothermic, occurs rapidly, and goes to completion when catalyzed by ruthenium on alumina pellets at 400°C. I first demonstrated a compact system appropriate for Mars application that united this reaction with a water electrolysis and recycle loop while working at Martin Marietta in Denver in 1993. The methane produced is great rocket fuel. The water can either be consumed as such or electrolyzed to make oxygen (for propellant or consumable purposes) and hydrogen (which is recycled).

Another system that has been demonstrated for Mars resource utilization is direct dissociation of CO_2 using zirconia electrolysis cells.

The reaction is:

$$CO_2 \Rightarrow CO + \tfrac{1}{2}O_2 \qquad (5.5)$$

Reaction 5.5 is very endothermic and requires the use of a ceramic system with high-temperature seals operating above 1,000°C. Its feasibility was first demonstrated by Robert Ash at the Jet Propulsion Lab in the late 1970s, and the performance of such systems has since been significantly improved by Kumar Ramohalli and K. R. Sridhar at the University of Arizona. (Sridhar has since formed a successful company, called Bloom Energy, devoted to commercializing this technology—operated in reverse as a kind of fuel cell—for use on Earth.) Its great advantage is that no cycling reagents are needed. Its disadvantage is that it requires a lot of power—about three times that of the Sabatier process to produce the same amount of propellant. A small-scale (producing twenty grams of oxygen per hour) version of such a system, called MOXIE, has been placed on the Mars 2020 rover, so we will soon get to see how well it works on Mars.[12]

Still another method of Mars propellant production is the reverse water-gas shift (RWGS).

$$CO_2 + H_2 \Rightarrow CO + H_2O \qquad (5.6)$$

This reaction is very mildly endothermic and has been known to chemistry since the nineteenth century. Its advantage over the Sabatier reaction is that all the hydrogen reacted goes into the water, from where it can be electrolyzed and used again, allowing a nearly infinite amount of oxygen to be produced from a small recycling hydrogen supply. It occurs rapidly at 400°C. However, its equilibrium constant is low, which means that it does not ordinarily go to completion, and it is in competition with the Sabatier reaction (5.4), which does. Working at Pioneer Astronautics in 1997, Brian Frankie, Tomoko Kito, and I demonstrated that copper on alumina catalyst was 100 percent specific for this reaction, however, and that by using a water condenser and air separation membrane in a recycle loop with a RWGS reactor, conversions approaching 100 percent could be readily achieved.

Our initial RWGS unit produced water at a rate of about one kilogram per day, which would be appropriate to make the oxygen propellant needed for the ascent vehicle of a robotic Mars sample return mission. Building on this work, in 2017, at Pioneer Energy, a commercial spin-off company of Pioneer Astronautics, my R&D team demonstrated an RWGS system operating at a rate of eighty kilograms of water production per day, sufficient to make all the oxygen propellant needed for the Mars Direct human exploration mission.

Running the RWGS with extra hydrogen, a waste gas stream consisting of CO and H_2 can be produced. This is known as "synthesis gas" and can be reacted exothermically in a second catalytic bed to produce methanol (reaction 5.7), propylene (reaction 5.8), or other fuels. Such use of RWGS "waste" gas to make methanol was first demonstrated during the 1997 Pioneer Astronautics program and then on a much larger scale (five kilograms of methanol per hour) during a 2017 Pioneer Energy project, while the propylene production reaction was demonstrated by the Pioneer Astronautics team during a program in 1998.

$$CO + 2H_2 \Rightarrow CH_3OH \qquad\qquad (5.7)$$

$$6CO + 3H_2 \Rightarrow C_3H_6 + 3CO_2 \qquad\qquad (5.8)$$

On Mars, buffer gas for breathing systems, consisting of nitrogen and argon, can be extracted directly from the atmosphere using pumps, as these gases comprise 2.7 percent and 1.6 percent of the air there, respectively. Water can also be extracted from the atmosphere using zeolite sorption beds, as shown by Adam Bruckner, Steve Coons, and John Williams at the University of Washington. Alternatively, it can be baked out of the soil, which the Mars Odyssey spacecraft has shown to vary from typically 5–10 percent water by weight at the equator to up to 60 percent in subarctic regions. Using ground-penetrating radar, subsurface liquid salt water has been found near the south pole by the European Mars Express spacecraft, while the NASA Mars Reconnaissance Orbiter has found massive formations of dust-covered glaciers made of nearly pure water ice in the northern hemisphere as far south as 38 degrees north. So relatively simple drilling, soil-baking, or ice-melting technologies should suffice to produce ample supplies of water on Mars.

Iron can also be produced on Mars very readily using either reaction 5.9 or 5.10. I say very readily because the solid feedstock, Fe_2O_3, is so omnipresent on Mars that it gives the planet its red color and thus, indirectly, its name.

$$Fe_2O_3 + 3H_2 \Rightarrow 2Fe + 3H_2O \qquad (5.9)$$

$$Fe_2O_3 + 3CO \Rightarrow 2Fe + 3CO_2 \qquad (5.10)$$

Reaction 5.9 is mildly endothermic (energy consuming) and can be used with a water electrolysis recycling system to produce oxygen as well. Reaction 5.10 is mildly exothermic (energy producing) and can be used in tandem with an electrolyzer and an RWGS unit to also produce oxygen. The iron can be used as such or turned into steel, since carbon, manganese, phosphorus, silicon, nickel, chromium, and vanadium, the key elements used in producing the principal carbon and stainless steel alloys, are all relatively common on Mars. To show this, in 2017, Pioneer Astronautics demonstrated the use of reaction 5.10 to make carbon steel out of Mars soil simulant samples.

The carbon monoxide produced by the RWGS can be used to produce carbon via:

$$2CO \Rightarrow CO_2 + C \qquad (5.11)$$

This reaction is exothermic and occurs spontaneously at high pressure and temperatures of about 600°C. The carbon so produced can be used to make carbon-carbon components or to produce silicon or aluminum via reactions 5.12 and 5.13.

$$SiO_2 + 2C \Rightarrow 2CO + Si \qquad (5.12)$$

$$Al_2O_3 + 3C \Rightarrow 2Al + 3CO \qquad (5.13)$$

Both SiO_2 and Al_2O_3 are common on Mars, so finding feedstock will be no problem. Reactions 5.12 and 5.13 are both highly endothermic, however. So except for specialty applications where aluminum is really required, steel will be the metal of choice for Martian construction.

On the Asteroids

The asteroids are rich in metals and also possess carbon, so much of the carbon-based resource utilization reactions developed for Mars could also be used there. Of special interest to asteroid miners will be a means of acquiring pure samples of various metals for purposes of commercial export. One way to do this is to produce carbonyls, as pointed out by the University of Arizona's Professor Lewis.[13]

For example, carbon monoxide can be combined with iron at 110°C to produce iron carbonyl ($Fe(CO)_5$), which is a liquid at room temperature. Then iron carbonyl can be poured into a mold and heated to about 200°C, at which time it will decompose. Pure iron, very strong, will be left in the mold, while the carbon monoxide will be released, allowing it to be used again. Similar carbonyls can be formed between carbon monoxide and nickel, chromium, osmium, iridium, ruthenium, rhenium, cobalt, and tungsten. Each of these carbonyls decomposes under slightly different conditions, allowing a mixture of metal carbonyls to be separated into its pure components by successive decomposition, one metal at a time.

An additional advantage of this technique is the opportunities it offers to enable precision low-temperature metal casting. You can take the iron carbonyl, for example, and deposit the iron in layers by decomposing carbonyl vapor, allowing hollow objects of any complex shape desired to be made. For this reason, carbonyl manufacturing and casting will no doubt also find extensive use on Mars and the thousands of worlds that lie just beyond.

The potential 3-D printing of these materials has only begun to be explored. But there is little doubt it can be done, and when it is, space settlers will be able to make anything they can draw.

Chapter 6

THE OUTER WORLDS

INTRODUCTION

The outer solar system is a vast arena thousands of times greater in extent than the sun's inner domain. It includes four spectacular giant planets, a minor planet, six moons of planetary size, and scores of smaller ones, as well as many known and possibly billions of unknown asteroidal and cometary objects of every description imaginable. It is a realm of stark beauty and unimaginable potential. Our first exploration probes have made amazing discoveries, including icy moons containing oceans dwarfing those of our home planet. Yet they have barely scratched the surface of the worlds of secrets nature has placed there. A generation of more capable spacecraft will be needed, with nuclear power to allow for active sensing through thick atmospheres and vastly increased data rates, and submarine vehicles to dive into ice-covered oceans to search for life. Yet still we will not know. The human mind will have to follow, and the challenge of the distances will demand the development of propulsion technologies far more capable than those needed for human missions to the moon or Mars. A measure of time thus will pass before the outer solar system becomes the domain of human activity, but it surely will. For, though the future can be but dimly seen, we already know that these outer worlds contain the keys to continued human survival, progress, and our posterity's hopes for the stars.

EXPLORATION OF THE OUTER SOLAR SYSTEM

Human exploration of the outer solar system began in a serious way with Galileo, who was the first to turn the telescope in that direction. The government of Venice rewarded Galileo handsomely for perfecting the spyglass, as it had great utility for naval warfare, but the church authorities were much less than pleased by what he saw in the heavens. Pointing his instrument into the night of January 7, 1610, Galileo identified the four major satellites (the "Galilean satellites") of Jupiter: Io, Europa, Ganymede, and Callisto, each comparable in size to the planet Mercury. This discovery of astronomical bodies orbiting a planet other than the Earth was a major blow to the Ptolemaic-Aristotelian worldview, upon which the church had pinned its authority. It also had extensive practical significance for the coming age of maritime discovery. By providing navigators with a completely reliable clock in the sky, the system of Galilean satellites allowed explorers to establish their longitude anywhere in the world (since if you can set your watch in agreement with some absolute standard, the time of sunrise will give you your longitude). Thus, by engaging in the apparently completely impractical activity of studying Jupiter, Galileo finally made it possible for humans to map and reliably navigate the Earth. The result was an age of long-range maritime commerce that generated fortunes that left the hoards of the Venetian merchant city-state to seem quite petty in comparison.

As telescopes grew and improved, other important finds followed, with the discovery of Saturn's rings and its moon Titan by Huygens in 1655; Jupiter's giant red spot by Cassini in 1665; the planet Uranus by Herschel in 1781; the planet Neptune by Adams and LeVerrier in 1846; Neptune's giant moon Triton by Lassell in 1846; Pluto by Tombaugh in 1930; Titan's atmosphere by Kuiper in 1944; and the giant iceteroid Chiron by Kowal in 1977. By the 1960s, Jupiter was known to have twelve satellites, Saturn nine, Uranus five, and Neptune two.

Outer solar system exploration was revolutionized in the 1970s with the advent of robotic exploration spacecraft, starting with the Pioneer 10 mission to Jupiter, which flew by the giant planet in 1972. This was followed by Pioneer 11 in 1973 and Voyager 1 in 1977, both of which made a close pass by Jupiter and then flew on to visit Saturn, reaching the ringed planet in 1979 and 1980, respectively. This was the first demonstration of multiple planet flyby missions using gravity assists, a technique that Voyager 2 took to a brilliant conclusion by visiting, in succession, Jupiter (1979), Saturn (1981), Uranus (1986), and Neptune (1989). The political maneuvers that made Voyager 2's mission possible were almost as tricky as the celestial mechanics. To save money, the Carter- and Reagan-era NASA headquarters brass, as well as bureaucrats in the Office of Management and Budget, wanted to limit Voyager 2's mission to Jupiter and Saturn only. To get the mission launched, the Jet Propulsion Lab's management had to assure them this would be the case, only later to gain agreement to return data from Uranus ("since the spacecraft was on its way there anyway") and then, pulling the same trick again, to survey Neptune as well.[1]

The Voyager missions were a tour de force and stand with Viking, Apollo, and Hubble as one of NASA's four greatest accomplishments to date. In addition to returning volumes of spectacular color images of the giant planets, the Voyagers imaged all the known moons and discovered literally dozens of additional ones. Ring systems around Jupiter, Uranus, and Neptune were discovered, as were many new features in Saturn's rings. Magnetic fields around the giant planets were measured, revealing an enormously powerful magnetic field and associated radiation belts circling Jupiter. Measurements were taken that showed the interior of the giant planets to be much warmer than anticipated. Both Voyagers actually imaged volcanoes in the process of erupting on Io, which was very unexpected, and which made abundantly clear the heating power of geothermal energy generated by tidal forces on bodies orbiting Jupiter. The importance of this was emphasized by another of Voyager's finds—namely, images of Europa showing the entire moon to be covered with ice. As revealed by Voyager, Europa's ice had fractures in it, suggesting to some a thick layer of sea ice lying above an ocean of liquid water. Prior to Voyager's observations of volcanoes on Io, no one would have thought that liquid water could possibly exist in the Jovian system. But if tidal forces could heat the interior of Io to the melting point of rock, was it not possible that the interior of Europa, the next major satellite outward from the giant planet, could be warmed tidally to above the melting point of ice? And if there were liquid water and heat beneath the ice of Europa, could that not potentially represent a home for life?

This speculation was amplified in 1996 and 1997, when the Galileo spacecraft imaged Europa again, at much finer resolution than Voyager, showing conclusively that Europa's ice covering is in fact sea ice, perhaps fifty kilometers thick, floating over an ocean of liquid water that is probably one hundred kilometers deep.[2] Not only is there liquid water on Europa, there is more liquid water on Europa than there is on Earth! During the 1980s, oceanographers had discovered deep-sea life subsisting on a food chain based not on photosynthesis but on chemosynthesis linked to hot vents on the seafloor. There seems to be no fundamental reason why similar ecosystems could not exist in the depths of Europa's ocean.

(See plate 10.)

Exploring Europa's ocean is now thus a major target for NASA's exobiological research program. The main problem is how one penetrates fifty kilometers of ice. At least at the surface, the ice is supercold—about −160°C. Ice at that temperature is as hard as rock. Even if human crews could be sent to Europa to set up drill rigs (which could be tough, as Europa is right in the middle of Jupiter's very dangerous radiation belts), drilling through that thickness of such material would be an incredible task. An alternative idea that I have suggested to NASA is that a radioisotope-heated sphere, built as strong as a cannonball, be released from a spacecraft and allowed to impact Europa's surface at high velocity. The sphere would thus bury itself beneath the ice and very slowly begin to melt its way down. The surface layer of meltwater around the probe would contain the chemicals, and perhaps frozen microbes, of the Europan ocean of the past. As the probe penetrated deeper, more recently created ice would be encountered and its water's captured contents made available for analysis. Thus, as the probe went deeper and deeper, it would produce a scan of Europa's ocean over a long period of geologic time. So long as the probe stayed in the ice, it could transmit data back to an orbiter using low-frequency radio. It could also potentially communicate using sound to a lander serving as a communication relay. If it reached the ocean, it would lose contact (since radio can penetrate ice much better than water) and sink rapidly, but it could still take measurements. Then, when it hit the bottom, ballast could be released, allowing the probe to float back up through the ocean, taking more measurements, until it hit the ice, when radio contact with the orbiter could be reestablished.

Since the probe would penetrate the ice quite slowly, perhaps only a few meters per day, it would be very advantageous to aim it for locations where the ice is thinnest, if we hope to reach to liquid ocean in a reasonable amount of time. This could be done by equipping the carrier spacecraft with long-wavelength ice-penetrating radar, which could map from orbit the thickness of the ice sheet covering Europa. If a thin region were identified, it could be then be targeted when the probe is released.

Figure 6.1. Artist's vision of the Cassini spacecraft at Saturn.
Image courtesy of NASA.

Such a mission would be quite ambitious, but a potentially easier route to searching for life in the ocean worlds of the outer solar system was opened up by the Cassini probe, which, launching from Earth in 1997, reached Saturn in 2004 and orbited among its moons, taking spectacular photographs of all of them, until it ran out of propellant in 2017 and was deorbited. One of the great discoveries of the Cassini mission was that Saturn's little moon Enceladus, only 255 kilometers in radius, is constantly spraying geysers of water hundreds of kilometers up into space from a region near its south pole.[3] Apparently, Enceladus is being slapped around by tidal interactions with Saturn on one side and its large moon Dione on the other, with the effect being both to heat pressurized subsurface water to hot liquid and to periodically crack the ice cover that is holding that water down. When the ice breaks, a Spindletop-like gusher occurs, with the difference being that the treasure being shot into the sky is not oil but underground ocean water that could potentially be gathered and sampled by an orbiting spacecraft for signs of life.

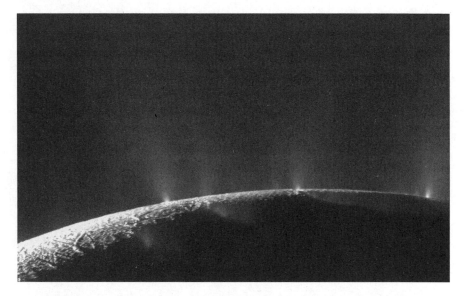

Figure 6.2. Photo taken by NASA's Cassini spacecraft of geysers spraying into space from cracks in the ice cover of Saturn's ocean moon Enceladus. If there is life in the ocean beneath the ice, these plumes could make it available for sampling. *Image courtesy of NASA.*

In the hope that such phenomena may also periodically occur on much easier-to-reach Europa—theory says that they should, and there is some inconclusive evidence that they do—NASA is planning a spacecraft called Europa Clipper to go into an elliptical orbit around Jupiter, where it would mostly avoid high radiation levels but still repeatedly encounter Europa to try to catch a whiff of a geyser.[4] If it is not so fortunate, it will still return a lot of data, as it will be heavily instrumented with all sorts of advanced cameras, spectrometers, magnetometers, thermal-emission-measuring instruments, and ice-penetrating radar. The spacecraft planned is so massive that it will need to launch on an SLS heavy-lift booster, which will not be available until at least 2022. In view of the fact that neither the Galileo nor Cassini orbiters needed a booster with more than a quarter of the SLS's lift capability to accomplish their missions, it would appear that an imperative to provide business for the SLS intruded on the design process. This sort of thinking is not helpful for program success and hopefully will be corrected.[5]

Be that as it may, in 2017 several competing plans for sending medium-sized spacecraft to orbit Enceladus—where we know the water plumes can be found—were submitted to NASA.[6] The proposing teams were very creative. With a little bit of luck, one of them should get a chance to go forward.

But the outer solar system is much more than a region of scientific interest. It could well prove to be the source of the resources needed to sustain an advancing human civilization—and to take us to the stars.

THE SOURCES OF POWER

In order to glimpse the probable nature of the human condition a century hence, it is first necessary for us to look at the trends of the past. The history of humanity's technological advance can be written as a history of ever-increasing energy utilization. If we consider the energy consumed not only in daily life but in transportation and the production of industrial and agricultural goods, then Americans in the electrified 1990s used approximately three times as much energy per capita as their predecessors of the steam-and-gaslight 1890s, who in turn had nearly triple the per capita energy consumption of those of the preindustrial 1790s. Some have decried this trend as a direct threat to the world's resources, but the fact of the matter is that such rising levels of energy consumption have historically correlated rather directly with rising living standard, and, if we compare living standards and per capita energy consumption of the advanced sector nations with those of the impoverished third world, continue to do so today. This relationship between energy consumption and the wealth of nations will place extreme demands upon our current set of available resources. In the first place, simply to raise the entire present world population to current American living standards (and in a world of global communications, it is doubtful that any other arrangement will be acceptable in the long run) would require increasing global energy consumption at least ten times. However, world population is increasing, and while global industrialization is slowing this trend, it is likely that terrestrial population levels will at least double before they stabilize. Finally, current American living standards and technology utilization are hardly likely to be the ultimate (after all, even in early twenty-first-century America, there is still plenty of poverty) and will be no more acceptable to our descendants a century hence than those of a century ago are to us. All in all, it is clear that the exponential rise in humanity's energy utilization will continue. In 2018, humanity mustered about twenty-three terawatts of power (one terawatt, TW, equals one million megawatts of power). At the current 2.6 percent rate of growth, we will be using nearly two hundred terawatts by the year 2100. The total anticipated power utilization and the cumulative resource used (starting in 2000) is given in table 6.1.

TABLE 6.1. PROJECTED HUMAN USE OF ENERGY RESOURCES

Year	Power	Energy used after 2000
2000	15 TW	0 TW-years
2025	28	516
2050	53	1,490
2075	101	3,350
2100	192	7,000
2125	365	13,700
2150	693	26,400
2175	1,320	50,600
2200	2,500	96,500

By way of comparison, the total known or estimated energy resources are given in table 6.2.

TABLE 6.2. SOLAR SYSTEM ENERGY RESOURCES

Resource	Amount
Known terrestrial fossil fuels	3,000 TW-years
Estimated unknown terrestrial fossil fuels	7,000
Nuclear fission without breeder reactors	300
Nuclear fission with breeder reactors	22,000
Fusion using lunar He3	10,000
Fusion using Jupiter He3	5,600,000,000
Fusion using Saturn He3	3,040,000,000
Fusion using Uranus He3	3,160,000,000
Fusion using Neptune He3	2,100,000,000

In table 6.2, the amount of He3 given for each of the giant planets is that present in their atmospheres down to a depth where the pressure is ten times that of the Earth's at sea level. If one extracted at a depth where the pressure was greater, the total available He3 would increase in proportion. If we compare the energy needs for a growing human civilization with the availability of resources, it is clear that even if the environmental problems associated with burning fossil fuels and nuclear fission are completely ignored, within a couple of centuries the energy stockpiles of the Earth and its moon will be effectively exhausted. Large-scale use of solar power can alter this picture somewhat, but sooner or later, the enormous reserves of energy available in the atmospheres of the giant planets must and will be brought into play.

Thermonuclear fusion reactors work by using magnetic fields to confine a plasma consisting of ultrahot charged particles within a vacuum chamber where they can collide and react. Since high-energy particles have the ability to gradually fight their way out of the magnetic trap, the reactor chamber must be of a certain minimum size so as to stall the particles' escape long enough for a reaction to occur. This minimum size requirement tends to make fusion power plants unattractive for low-power applications, but in the world of the future, where human energy needs will be on a scale tens or hundreds of times greater than today, fusion will be far and away the cheapest game in town.

A century or so from now, nuclear fusion using the clean-burning (no radioactive waste) deuterium-helium-3 reaction will be one of humanity's primary sources of energy, and the outer planets will be the Persian Gulf of the solar system.

THE PERSIAN GULF OF THE SOLAR SYSTEM

Today the Earth's economy thirsts for oil, which is transported over oceans from the Persian Gulf and Alaska's North Slope by fleets of oil-powered tankers. In the future, the inhabitants of the inner solar system will have the fuel for their fusion reactors delivered from the outer worlds by fleets of spacecraft driven by the same thermonuclear power source. For while the ballistic interplanetary trajectories made possible by chemical or nuclear thermal propulsion are adequate for human exploration of the inner solar system and unmanned probes beyond, something a lot faster is going to be needed to sustain interplanetary commerce encompassing the gas giants.

Fusion reactors powered by D-He3 are a good candidate for a very advanced spacecraft propulsion. The fuel has the highest energy-to-mass ratio of any substance found in nature, and, further, in space the vacuum that the reaction needs to run in can be had for free in any size desired. A rocket engine based upon controlled fusion could work simply by allowing the plasma to leak out of one end of the magnetic trap, adding ordinary hydrogen to the leaked plasma, and then directing the exhaust mixture away from the ship with a magnetic nozzle. The more hydrogen added, the higher the thrust (since you're adding mass to the flow), but the lower the exhaust velocity (because the added hydrogen tends to cool the flow a bit—if you are interested in the math, see note).[7] For travel to the outer solar system, the exhaust would be more than 95 percent ordinary hydrogen, and the exhaust velocity would be more than 250 km/s (a specific impulse of 25,000 seconds, which compares quite well with the specific impulses of chemical or nuclear thermal rockets of 450 or 900 seconds respectively). Large nuclear electric propulsion (NEP) systems using fission reactors and ion engines, a more near-term possibility than fusion, could also achieve 25,000-second specific impulse. However, because of the complex electric conversion systems such NEP engines require, the engines would probably weigh an order of magnitude more than fusion systems, and as a result, the trips would take about twice as long. If no hydrogen is added, a fusion configuration could theoretically yield exhaust velocities as high as 15,000 km/s, or 5 percent the speed of light! Although the thrust level of such a pure D-He3 rocket would be too low for in-system travel, the terrific exhaust velocity would make possible voyages to nearby stars with trip times of less than a century.

Extracting the He3 from the atmospheres of the giant planets will be difficult, but not impossible. What is required is a winged transatmospheric vehicle that can use a planet's atmosphere for propellant, heating it in a nuclear reactor to produce thrust. I call such a craft a NIFT (for Nuclear Indigenous Fueled Transatmospheric vehicle). After sortieing from its base on one of the planet's moons, a NIFT would either cruise the atmosphere of a gas giant, separating out the He3, or rendezvous in the atmosphere with an aerostat station that had already produced a shipment. In either case, after acquiring its cargo, the NIFT would fuel itself with liquid hydrogen extracted from the planet's air, then rocket out of the atmosphere to deliver the He3 shipment to an orbiting fusion-powered tanker bound for the inner solar system.

Table 6.3 shows the basic facts that will govern commerce in He3 from the outer solar system. Flight times given are one-way from Earth to the planet, with the ballistic flight times shown being those for minimum-energy orbit transfers. These can be shortened somewhat at the expense of propellant (gravity assists can help too but are available too infrequently to support regular commerce) but, in any case, are too long for commercial traffic to Saturn and beyond; even if the vessels are fully automated, time is money. The NEP and fusion trip times shown assume that 40 percent of the ship's initial mass in Earth orbit is payload, 36 percent is propellant (for one-way travel; the ships refuel with local hydrogen at the outer planet), and 24 percent is engine. Jupiter is much closer than the other giants, but its gravity is so large that even with the help of its very high equatorial rotational velocity, the velocity required to achieve orbit is an enormous 29.5 km/s. A NIFT is basically a nuclear thermal rocket (NTR) with an exhaust velocity of about 9 km/s, and so even assuming a "running start" airspeed of 1 km/s, the mass ratio it would need to achieve such an ascent is more than twenty. This essentially means that Jupiter is off-limits for He3 mining, because it's probably not possible to build a hydrogen-fueled rocket with a mass ratio greater than six or seven. On the other hand, with the help of lower gravity and still-large equatorial rotational velocities, NIFTs with buildable mass ratios of about four would be able to achieve orbit around Saturn, Uranus, or Neptune.

TABLE 6.3. GETTING AROUND THE OUTER SOLAR SYSTEM

Planet	Distance from sun		One-way flight time		Velocity to orbit	NIFT mass ratio
		Ballistic	NEP	Fusion		
Jupiter	5.2 AU	2.7 years	2.2 years	1.1 years	29.5 km/s	23.7
Saturn	9.5	6.0	3.0	1.5	14.8	4.6
Uranus	19.2	16.0	5.0	2.5	12.6	3.6
Neptune	30.1	30.7	6.6	3.3	14.2	4.3

TITAN

As Saturn is the closest of the outer planets whose He3 supplies are accessible to extraction, it will most likely be the first of the outer planets to be developed. The case for Saturn is further enhanced by the fact that the ringed planet possesses an excellent system of satellites, including Titan, a moon which, with a radius of nearly 2,600 kilometers, is actually larger than the planet Mercury.

It's not just size that makes Titan interesting. Saturn's largest moon possesses an abundance of all the elements necessary to support life. It is believed by many scientists that Titan's chemistry may resemble that of the Earth during the period of the origin of life, frozen in time by the slow rate of chemical reactions in a low-temperature environment. These abundant pre-biotic organic compounds comprising Titan's surface, atmosphere, and oceans can provide the basis for extensive human settlement to support the Saturnian He3 acquisition operations.

Because of its thick, cloudy atmosphere, the surface of Titan is not visible from space, and many basic facts about this world remain a mystery. Here's what we know.

Titan's atmosphere is composed of 90 percent nitrogen, 6 percent methane, and 4 percent argon. The atmospheric pressure is 1.5 that of Earth sea level, but because of the surface temperature of 100 K (–173°C), the density is 4.5 Earth sea level. The surface gravity is one-seventh that of the Earth, and the wind conditions are believed to be light.

The latest evidence from the Cassini orbiter and its Huygens lander revealed that that surface consists of a mixed terrain, including deserts with dunes of water ice crystals and rocky mountains in the tropics; polar regions filled with networks of liquid methane streams, rivers, lakes, and seas; and midlatitude regions featuring both dry land and some made swamplike by its liquid methane moisture content.

The same nuclear thermal rocket engines that power NIFT vehicles mining Saturn's atmosphere could employ the methane abundant in Titan's atmosphere as propellant to enable travel not only all over Titan but throughout most of the Saturnian system. For example, because of Titan's thick atmosphere and low gravity, an eight-ton nuclear thermal flight vehicle operating in an air-breathing mode in Titan's atmosphere at a flight speed of 160 km/hr would require a wing area of only four square meters to stay aloft—in other words, virtually no wings at all. Employing the methane as rocket propellant in an NTR engine, a specific impulse of about 560 seconds (5.5 km/s exhaust velocity) could be achieved. The ΔV required to takeoff from Titan and go onto an elliptical orbit with a minimum altitude just above Saturn is only 3.2 km/s. Because the specific impulse of the rocket is high and the required mission ΔV is low, the mass ratio of the Titan-Saturn NTR ferry would only have to be about 1.8, which means that it could deliver a great deal of cargo. The downward-shipped cargo would be released in pods equipped with aeroshields that would allow them to brake from the elliptical transfer orbit down to the low circular orbit of a Saturn helium-3 processing station, which supports the operation of the Saturn-diving NIFTs. After releasing the cargo pods, the ferry would continue on its elliptical orbit until it reached its apogee at Titan's distance from Saturn, just six days after its initial departure. Because Titan's orbital period is sixteen days, it would not be there to meet the ferry. So a small rocket burn would be effected that would raise the orbit's periapsis (low point) a bit, thereby adjusting the orbital period of the ferry to ten days, allowing it to rendezvous with Titan and aerobrake and land on the next go round. Most of the cargo delivered to low Saturn would be supplies or crew for the orbiting NIFT base. However, some could be pods filled with methane propellant. These could be stockpiled at the orbiting station. When enough are accumulated to enable the 9 km/s ΔV needed to travel from low Saturn orbit onto a trans-Titan trajectory, a ferry could aerobrake itself and go to the station, and then be used to ship crew or cargo back to Titan.

Alternatively, it might also be found desirable to use some of Saturn's lower moons (several of which are quite sizable and may represent developable worlds in their own right) as intermediate bases. Nearly all of these have substantial water ice deposits, making fusion thermal steam rockets of the type used in the main asteroid belt attractive for internal transportation in the Saturn system. This could make ferry operations a lot easier.

The propulsion requirements to travel from Titan to Saturn's larger moons (counting the small fry, there are at least sixty-two in all) are shown in table 6.4. Each excursion involves landing on the destination moon twice, engaging in activity at two locations separated by up to forty degrees of latitude or longitude, and then returning to aerobrake and refuel at Titan.

TABLE 6.4. TITAN-BASED, METHANE-PROPELLED NTR EXCURSIONS TO SATURN'S OTHER SATELLITES

Destination	Distance from Saturn	Radius	ΔV	Mass ratio
Mimas	185,600 km	195 km	13.17 km/s	11.0
Enceladus	238,100	255	11.25	7.77
Tethys	294,700	525	10.05	6.24
Dione	377,500	560	8.60	4.79
Rhea	527,200	765	6.91	3.52
Titan	1,221,600	2575	0.00	1.00
Hyperion	1,483,000	143	3.84	2.01
Iapetus	3,560,100	720	6.90	3.52
Phoebe	12,950,000	100	8.33	4.56

Since methane is more than six times as dense as hydrogen, NTR vehicles using methane propellant should be able to achieve mass ratios greater than eight. It can be seen that with such capability, Titan-based NTR vehicles will be able to travel to and from all of Saturn's moons, except Mimas, virtually at will. If the vehicle can refuel on the destination moon, for example by using water or hydrogen from local ice as propellant for the return flight, the required mass ratios would shrink to the square root of those shown. Thus, if we had refueling bases on both Titan and Enceladus, the mass ratio needed to travel between them would be only 2.8, enabling the transport of large cargos.

In certain ways, Titan is the most hospitable extraterrestrial world within our solar system for human colonization. In the almost Earth-normal atmospheric pressure of Titan, you would not need a pressure suit, just a dry suit to keep out the cold. On your back, you could carry a tank of liquid oxygen, which would need no refrigeration in Titan's environment, would weigh almost nothing, and which could supply your breathing needs for a weeklong trip outside of the settlement. A small bleed valve off the tank would allow a trickle of oxygen to burn against the methane atmosphere, heating your breathing air and suit to desirable temperatures. With one-seventh Earth gravity and 4.5 times the atmospheric density of terrestrial sea level, humans on Titan would be able to strap on wings and fly like birds! (Just as in the story of Daedalus and Icarus—though being more than nine times distant from the sun than from Earth, such fliers wouldn't have the worry of their wings melting). Electricity could be produced in great abundance, as the 100-K heat sink available in Titan's atmosphere would allow for easy conversion of thermal energy from nuclear fission or fusion reactors to electricity at efficiencies of better than 80 percent. Most important, Titan contains billions of tons of easily accessible carbon, hydrogen, nitrogen, and oxygen. By utilizing these elements together with heat and light from large-scale nuclear fusion reactors, adding seeds and some breeding pairs of livestock from Earth, a sizable agricultural base could be created within a protected biosphere on Titan.

COLONIZING THE JOVIAN SYSTEM

We have discussed colonization of Saturn and the major planets beyond. Why not Jupiter, which is much closer to Earth and has four giant moons to Saturn's one? The answer is that as interesting as Jupiter's system is scientifically, its development will probably follow that of Saturn, primarily because the giant planet's enormous gravitational field makes extracting its atmospheric helium-3 supplies extremely difficult. Another problem facing the development of Jupiter is its extremely powerful radiation belts, within which many of its moons orbit.

The following table shows the primary moons of Jupiter's system (there are at least sixty-seven altogether) and gives the radiation dose that would be experienced by an unshielded human on the surface of each. The radiation doses are my own calculation, based upon data produced by James Van Allen during the Pioneer 10 and 11 missions.

TABLE 6.5. THE JUPITER SYSTEM

Moon	Distance from Jupiter	Radius	Radiation dose
Metis	127,960 km	20 km	18,000 rem/day
Adrastea	128,980	10	18,000
Amalthea	181,300	105	18,000
Thebe	221,900	50	18,000
Io	421,600	1,815	3,600
Europa	670,900	1,569	540
Ganymede	1,070,000	2,631	8
Callisto	1,883,000	2,400	0.01
Leda	11,094,000	8	0
Himalia	11,480,000	90	0
Lysithea	11,720,000	20	0
Elara	11,737,000	40	0
Ananke	21,200,000	15	0
Carme	22,600,000	22	0
Pasiphae	23,500,000	35	0
Sinope	23,700,000	20	0

In table 6.5, radiation doses of "0" mean negligible doses from Jupiter's radiation belts as such. There still would be the normal cosmic ray doses of about 0.14 rem/day. Also, while doses from Jupiter's belts would be negligible under normal circumstances, they would be much higher when the satellites occasionally pass through Jupiter's enormous magnetotail, which extends in the antisunward direction from Jupiter for hundreds of millions of kilometers. Presumably, however, people could take cover underground during these occurrences.

A radiation dose of 75 rem or more, if delivered during a time that is short compared to the cell repair and replacement cycles of the human body—say, thirty days—will generally cause radiation sickness, while doses of more than 500 rem will result in death. On Europa and all moons further in, such fatal doses would be administered to unshielded humans within a single day. On Ganymede, the dose rate is not too bad, provided that people would generally stay in shielded quarters (which they could readily build out of very strong 120-K ice) and only come out on the surface for a few hours now and then to perform essential tasks. On Callisto and those moons further out, Jupiter's radiation belts are not an issue, except during the time of magnetotail pass-through, as discussed above.

So, of Jupiter's planetary-sized satellites, only Callisto and perhaps Ganymede can be considered reasonable targets for human settlement. They are big places and possess such necessary elements as water, carbonaceous material, metals, and silicates. Sunlight is too dim for solar to be an attractive power supply (although NASA's solar-powered Juno probe has shown that it is possible), but there is a reasonable chance that geothermal power generated by tidal interaction with Jupiter may well be available on Ganymede (it certainly is on Io, the innermost major moon, whose active volcanoes are so numerous that some were photographed during eruption by the Voyager probe) and possibly Callisto as well. The moons beyond Callisto are probably captured asteroids. Their main attraction compared to those in the Main Belt would be that they are permanently stationed in the Jupiter system—if a significant branch of human society should develop on Callisto, those moons' exploitation could readily be supported from that location.

Jupiter's curse is its gravity field. Paradoxically, that may also prove to be its greatest resource. Jupiter is unmatched in its ability to sling-shot a spacecraft on an exceptionally fast trajectory with no propel-lant costs. You simply "drop" your spacecraft so it falls toward Jupiter, but misses, to perform a fast swing by instead. If your spacecraft is not bound to Jupiter—say it is heading into the outer solar system from Earth—this is easy to do. By whipping past Jupiter, you can use such a "gravity assist" to add a great deal of velocity to the spacecraft at no cost in propellant. This was the trick that enabled the Voyager missions.

But even if you are on an orbit that is bound within Jupiter's system, you can still use its gravity to generate fast departure veloci-ties. I know this statement sounds bizarre, especially to people who know their basic physics, but it's true!

Let's say you are living on the manufacturing colony of Callisto and you want to ship out some supplies fast to the helium-3 mining operation based on Saturn's moon Titan. There's ice on Callisto, so you can use this to make hydrogen/oxygen rocket propellant to get to orbit and perform high-thrust maneuvers in the Jupiter system. It will require a ΔV of about 2.4 km/s to take off from Callisto and reach a highly elliptical parking orbit about the moon. There you refuel, or transfer yourself and some propellant to a dedicated interplanetary spacecraft, and then execute a ΔV of 1.4 km/s to depart Callisto onto an elliptical orbit with its closest approach to Jupiter at 489,000 kilo-meters from the planet's center. This orbit will have a period exactly half that of Callisto's, so after two of your orbits, you will meet Callisto again (16.7 days later). Along the way, you make it your business to pass close by either Europa or Ganymede, and you use their gravity to distort your orbit a bit, so as to give you an increased encounter velocity when you return to Callisto. At that point, you perform still another gravity assist to lower your closest approach to Jupiter to 78,640 kilometers from the giant's center, which means you will pass above its surface at an altitude of 7,150 kilometers. This will take you through the thick of Jupiter's radiation belts, and any crew or sensi-tive electronics aboard will have to be well shielded. Because you have dived so low, your velocity at minimum altitude will be an enormous

55.7 km/s (nearly 125,000 mph). Jupiter's escape velocity at that altitude is 56.8 km/s, so a little extra push of 1.1 km/s would allow you to depart into interplanetary space. But instead of giving a little push, you give a big one, firing your chemical rocket to deliver a ΔV of, say, 6 km/s. Rocket propulsion systems don't know or care how fast you are flying; they only know how much velocity they add. But the energy of a spacecraft is a function of the square of its velocity. So, the faster you are already going, the more energy you add to the trajectory with a given velocity addition. The relevant equation is:

$$V_d^2 = V_{max}^2 - V_e^2 \qquad (6.1)$$

Where V_d is the velocity that the spacecraft departs the planet, V_{max} is the maximum velocity achieved right after the spacecraft fires its engine during its fast dive through low orbit, and V_e is the planet's escape velocity at the lowest point of the orbit. Without going into the math here, let's show the effect of applying it to the case at hand, firing our rocket engine during the above-described low pass over Jupiter. The results are shown in table 6.6.

TABLE 6.6. DEPARTING JUPITER AT HIGH VELOCITY USING HIGH-THRUST ROCKETS
(INITIAL ORBIT IS 78,640 KM × 1,883,000 KM AROUND JUPITER'S CENTER)

Rocket ΔV (km/s)	Maximum velocity (km/s)	Departure velocity (km/s)
1.1	56.8	0
1.5	57.2	6.8
2	57.7	10.2
3	58.7	14.8
4	59.7	18.4
5	60.7	21.4
6	61.7	24.1
7	62.7	26.6
8	63.7	28.8
9	64.7	31.0
10	65.7	33.0

So, in exchange for your rocket's own velocity increment of 6 km/s imparted during your orbital dive, you can go screaming out of the Jupiter system at the phenomenal clip of 24 km/s! That's an initial speed for your spacecraft of nearly 5 AU per year, sufficient to get you out to Saturn or back to Earth in less than one year, or to Uranus in three years, with nothing but chemical propulsion. Advanced propulsion systems on board, such as nuclear electric propulsion or fusion, could then be used to accelerate the system even more subsequent to departure.

Thus, once there is helium-3 commerce to be supported in the outer solar system, Jupiter, using the resources of its outer moons and its gravity well, could develop as an important solar system transportation node.

Nineteenth-century New Englanders thought they had an unmatchable racket going selling ice. Imagine the envy of those sharp-minded old-time Yankees if they could awake from their graves and look into the future to see Callisto colonists selling . . . gravity!

THE KUIPER BELT AND OORT CLOUD

> *It is generally considered that beyond the Sun's family of planets there is absolute emptiness extending for light-years until you come to another star. In fact it is likely that the space around the Solar System is populated by huge numbers of comets, small worlds a few miles in diameter, rich in water and other chemicals essential to life. . . . Comets, not planets, are the major potential habitat of life in space.*
>
> —Freeman Dyson, 1972

As mentioned earlier, beyond Neptune lie two zones of asteroid-sized objects rich in volatiles. The innermost such region is the Kuiper Belt. Consisting of millions of iceteroids orbiting more or less in the same plane as the planets ("the ecliptic"), it begins about 30 AU and extends to perhaps 50 AU. There then appears to be a gap largely free of such objects, until at about 1,000 AU a new iceteroid zone begins. This is the Oort cloud, whose trillions of frozen objects populate a sphere surrounding our solar system in all directions out to about 100,000 AU—roughly halfway to the nearest star. Because they are so far from the sun, such objects orbit very slowly. At 10,000 AU, for example, the speed required to orbit the sun is just 300 m/s (compared to the Earth's 30,000 m/s), so it only takes a very mild velocity change to radically perturb such objects' orbits. It is believed that such pertur-bations occasionally occur naturally, when passing stars pass through the Oort cloud and disturb its orbits with their gravitational fields. When that happens, one or more of the iceteroids (from either our Oort cloud or that of the visiting star, which our sun will be disrupting as well) can be displaced from their peaceful existence in the outer darkness. As they fall toward the sun, they speed up enormously, and with volatiles boiling away, they come blazing into the inner solar system as gigantic young comets.

Such comets have sometimes hit Earth, with effects similar to those resulting from asteroidal impacts. However, unlike near-Earth asteroids, which spend their lives in the inner solar system and which can, in principle, be spotted and have their trajectories mapped many orbits before a potential Earth-smashing collision, comets can emerge from the dark and come in fast and hard with the advantage of sur-prise. The only way to control them is to detect and deflect them while they are still very far out. This means that someday, for security pur-poses if no other, there will be a need for a substantial human presence and technical capability in both the Kuiper Belt and the Oort cloud.

But there may be other reasons that drive humans to populate this vast archipelago of cosmic islands. Based on analysis of comets, it's fairly clear that the volatile iceteroids of the Oort cloud are rich not just in water but in carbon and nitrogen, much of it in the form of the usual compounds of organic chemistry and life. In addition, some of the most essential elements of industry, including iron, silicon, magne-sium, sulfur, nickel, and chromium, are present in modest but possibly sufficient concentrations. This has caused some, notably the visionary Princeton professor Freeman Dyson, to identify these bodies as a major arena for the human future.[8]

It's rather futuristic, but not impossible. The inhabitants of such places wouldn't really need much steel for their constructions. For most purposes, ice—lightweight, cheap, and superstrong at 20 kelvin (−253°C)—would serve quite well. Incredible degrees of both robotic automation and human versatility will be required to compensate for the limited division of labor possible in such small, widely scattered colonies, but perhaps earlier human experience in coping with a lesser degree of this same problem while settling the asteroid belt will pave the way. The main missing ingredient is energy. While some have suggested concentrating starlight, it doesn't really make sense. To get a single megawatt of power, the mirror would have to be the size of the continental United States. The only viable alternative based on currently known physics is fusion. In the Kuiper Belt, it might be possible to get helium-3 shipped out from mining operations around Neptune. Oort cloud settlements would be too far out to obtain much from the solar system, though deuterium should be available in all ice-teroids, so perhaps the colonists might choose to build reactors based on that fuel alone. However, helium can exist in the liquid phase below 5 kelvin (−268°C), which is the environmental temperature at about 3,000 AU. It is therefore not impossible that liquid helium could exist within Oort cloud objects beyond that distance. Helium is the second most abundant element in the cosmos, and at very low temperatures, it could accrete within hydrogen ice objects into a helium-rich ice-teroid. For Oort cloud colonists, such an object would be quite a find!

Perhaps the same wanderlust and reach for diversity that drove the old folks to settle the asteroid Main Belt in the twenty-first century will move their descendants a century or more later to try their luck among the million untamed worlds of the Kuiper frontier. Why go? Why stay? Why live on a planet whose social laws and possibilities were defined by generations long dead, when you can be a pioneer and help to shape a new world according to reason as you see it? The need to create is fundamental. Once started, the outward movement will not stop.

THE ROAD TO THE STARS

The two main obstacles to settling the outer solar system are power and transportation. As mentioned earlier, solar energy in the realm of the gas giants and beyond is negligible. However, in the era we are discussing, we can expect that fusion powered by helium-3 will be the dominant energy source. Indeed, the need to acquire helium-3 to fuel such systems will be one of the prime motivations for the colonization of the far worlds of the outer solar system.

As to the issue of transportation, well, I call current space transportation systems First Generation. These are sufficient for launch into Earth orbit; for manned missions to the moon, Mars, and near-Earth asteroids; and for limited-capability unmanned probes to other planets. For colonization of the inner solar system, out to the Main Belt, we need to move on to Second Generation systems, typified by nuclear thermal rocket propulsion, nuclear electric propulsion, and advanced aerobraking technology. Such Second Generation systems also open capabilities for vastly expanded unmanned exploration of the outer solar system. They are, however, marginal for manned colonization of Titan, as the three- to four-year one-way flight times they impose upon this mission are excessive. However, as the fusion economy initiated by the moon's supply of He3 grows, demands will be developed that can only be satisfied by the vastly larger stocks of this substance available in the outer solar system. By improving the in-space life support systems associated with Second Generation, a few pioneers will make their way to Titan using Second Generation transportation technologies. Once even a small base is established on Titan, there will be a tremendous incentive to develop Third Generation systems, such as fusion propulsion (especially since we will then have the abundant He3 supplies needed to fuel them). This will allow for quick trips and rapid development of Titan and the rest of the outer solar system. Such Third Generation propulsion systems, however, together with fully Third Generation closed-cycle ecological life support, will enable travel beyond the nine known planets to the Oort cloud and, when advanced to their limits, create a basis for interstellar missions, with flight times to nearby stars on the order of fifty to a hundred years.

Humans will go to the outer solar system not merely to work but to live, to love, to build, and to stay. But the irony of the life of pioneers is that if they are successful, they conquer the frontier that is their only true home, and a frontier conquered is a frontier destroyed. For the best of humanity, then, the move must be ever outward. The farther we go, the farther we will become able to go, and the farther we shall need to go. Ultimately, the outer solar system will simply be a way station toward the vaster universe beyond. Just as Columbus's discovery of the New World called into being the full rigged sailing ships, steamers, and Boeing 707s that allowed the rest of humanity to follow in his wake, so those brave souls who dare the great void to our neighbor stars with ships of the Third Generation will draw after them a set of Fourth Generation space transportation systems, whose capabilities will open up the galaxy for humankind.

For while the stars may be distant, human creativity is infinite.

FOCUS SECTION: THE ENTREPRENEURIAL FUSION REVOLUTION

Human expansion to the outer solar system and then the stars depends critically on the development of thermonuclear fusion power. The vast supplies of He3 on the moon and in the atmospheres of the outer planets will only become a useful energy resource if fusion reactors are developed. Only fusion offers the power source needed by a growing Type II civilization to rebuild asteroids into city-states or terraform sterile planets into living worlds. And only fusion energy, whether employed directly in the form of fusion rockets or transformed into antimatter or laser light, can provide the enormous amounts of power required for interstellar travel on human timescales.

So if humans are to have a future among the stars, we need fusion power. But, after achieving considerable progress from the 1950s through the 1980s, the world's government-backed fusion power research programs have been mired in nearly complete stagnation for the past quarter century. Is the situation hopeless?

No, it is not. The national fusion programs progressed well during the Cold War because of fierce international competition. They have stopped moving forward since the late 1980s because the decision to consolidate them all into a single global project, the International Thermonuclear Experimental Reactor (ITER), removed all stimulus for action. Indeed, it took nearly a quarter century for the bureaucrats in charge of ITER to manage to reach a consensus in 2010 on where to put it, and it will be another quarter century before the machine even attempts to reach thermonuclear ignition in 2035.

This absurdly glacial rate of advance caused many people in and around the technical community to become cynical. "Fusion is the energy of the future, and always will be" became a common quip.

But then a breakthrough happened. SpaceX demonstrated that it is possible for a well-run, lean, and creative entrepreneurial organization to achieve things—and do so much more quickly—that were previously thought to require the efforts of governments of major powers. This hit observers of the fusion program like a bolt from the blue. Could it be that the seemingly insurmountable barriers to the achievement of controlled fusion—like the barriers to the attainment of cheap space launch—were not really technical, but institutional? Venturesome investors suddenly became interested.

I worked at Los Alamos in 1985 as a graduate student intern on a then-novel fusion concept called the spherical tokamak (ST). I can remember one lunch when our group leader, Robert Krakowski, philosophically told the rest of us, "You know, when fusion power is finally developed, it won't be at a place like Los Alamos or Livermore. It will be by a couple of crackpots working in a garage."

We laughed at that then. Maybe Krakowski went a bit too far. But if not a couple of crackpots working in a garage, how about an A-team of crack engineers working in a warehouse?

It's A-team time. As a result of Musk's shot heard 'round the world, a whole raft of innovative private fusion power start-ups are getting funded. Here's a bit about some of them.

1. Tokamak Energy. This Oxfordshire, England, venture, started in 2009 by former Culham Laboratory staffers Jonathan Carling, David Kingham, and Michael Graznevitch, has raised $50 million of mostly private money to try to develop the ST (the same

concept that I worked on in the 1980s, which was too innovative for ITER to adopt) into a commercial reactor. In a magnetic confinement fusion reactor, the amount of power that can be generated rises in proportion to $\beta^2 B^4$, where β is the ratio of the plasma pressure to the magnetic pressure, and B is the magnetic field strength. An ordinary tokamak like ITER can only achieve a β of about 0.12, but an ST can achieve a β of 0.4. As a result, an ST can produce the same amount of power as a regular tokamak in a machine less than one-tenth the size and cost.

Figure 6.3. ITER under construction (*left*); Tokamak Energy's Spherical Tokamak (*right*). *Image courtesy of ITER and Tokamak Energy; reproduced by permission, © 2018 by Tokamak Energy Ltd.*

2. Commonwealth Fusion Systems. Founded in 2018, this MIT-based venture has raised $75 million so far, including $50 million from the Italian oil company ENI and about $25 million from the Breakthrough Energy Ventures fund backed by Bill Gates, Jeff Bezos, Jack Ma, Mukesh Ambani, and Richard Branson. The root of the CFS design concept goes back to the 1980s, when the very creative maverick MIT physicist Bruno Coppi proposed achieving fusion in a very small tokamak by the simple expedient of using ultrastrong magnetic fields. Tokamaks are toroidal chambers with a magnetic field running the long way around the doughnut. The magnetic field lines confine particles to follow them, spiraling around the chamber, with the radius of the spirals being inversely

proportional to the strength of the magnetic field. Coppi reasoned that the relevant dimension of a tokamak was not its size per se but the ratio of its size to the radius of the spiral, because it is this ratio that determines how long a particle will last before it hits the wall. Furthermore, as noted above, the higher the magnetic field strength, the more quickly the particle is likely to react. So if you want a particle to take part in a fusion reaction before it hits the wall (which would cool it too much for fusion), the key is just to go for broke with ultrapowerful magnets. But the problem is that the highest magnetic field it is practical to achieve with traditional low-temperature superconducting magnets is about six tesla, and Coppi needed twelve. So he designed an experimental machine called "Ignitor" using copper magnets. This could not be practical as a commercial reactor because the resistive copper magnets would use too much power. Nevertheless, if it had been built, we probably would have achieved thermonuclear fusion ignition in the 1990s. But all of the US Department of Energy funds were committed to ITER, so Ignitor was never built. However, starting around 2014, an MIT group led by Professor Dennis Whyte decided to pick up where Coppi had left off, improving on the Ignitor concept by making use of high-temperature superconductor magnets, which require no electric power and can reach twelve tesla. As a result, with more than twice the magnetic field strength as ITER, the CFS reactor, known as SPARC (for Smallest Possible Affordable Robust Compact) fusion reactor, will achieve one-fifth the power hoped for by ITER in a reactor a sixty-fifth the volume. Furthermore, CFS aims to do it by 2025, achieving in seven years what ITER hopes to do in half a century.

3. Tri Alpha Energy. Founded in 1998 by the late Dr. Norman Rostoker, southern California-based TAE recently received more than $500 million in investment from heavy hitters including Microsoft cofounder Paul Allen, Goldman Sachs, Wellcome Trust, Silicon Valley's New Enterprise Associates, and Venrock. TAE's departure from orthodoxy is more radical than the abovementioned start-up in that they do not use a tokamak or toroidal chamber of any kind. Instead, TAE uses a simple cylinder chamber, with the required

toroidal magnetic field induced in the plasma itself by having a linear magnetic field created by an outside solenoid suddenly reversed, causing it to curve around and connect to itself. This creates a kind of smoke-ring-current vortex in the plasma, or what is called in the fusion business a "field-reversed configuration" or FRC. When I was in graduate school in the 1980s at the University of Washington, FRCs were all the rage, as they routinely achieve β values of more than 0.5. Moreover, their simple construction makes them potentially much more promising for creating low-cost commercial systems or fusion rocket drives than tokamaks. But by the 1980s, tokamaks had crowded out all funding within the US fusion budget, and shortly afterward, even the US tokamaks were starved for funds to feed ITER. FRCs were far too avant-garde to even be considered by ITER. But private investors are much more daring than international bureaucrats, and TAE is pushing hard, with a goal of demonstrating net energy production by 2024.

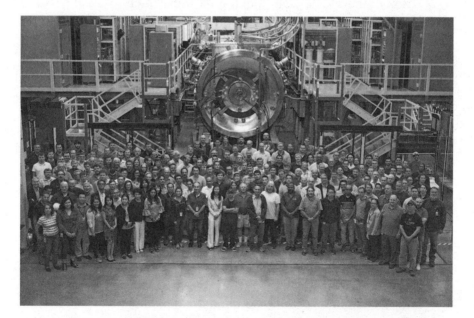

Figure 6.4. Tri Alpha Energy has received more than $500 million in private investment to develop the field-reversed configuration into a practical fusion reactor. *Image courtesy of TAE Technologies.*

4. Helion Energy. Founded in 2005 by Professor John Slough of the University of Washington, Helion uses two FRCs that are accelerated into a cylindrical reaction chamber from opposite ends to collide in the middle, where they are compressed by a solenoidal magnetic field to reaction conditions. Fusion reactions then heat the FRC plasma, causing it to expand back toward the chamber ends at high speed, with its energy being directly converted into electricity in the process. The cycle would then be repeated once per second to keep making power (or, alternatively, rocket thrust). Helion recently got an investment of $14 million from Peter Thiel's Mithril Fund.

5. General Fusion. Founded in Burnaby, British Columbia, by Dr. Michel Laberge and Michael Delage in 2002, GF has since received some $130 million in investment. The GF concept injects an FRC into a chamber containing a rotating liquid metal wall, which is then driven inward by an array of pistons to compress the FRC to fusion conditions. This is a variant of the "imploding liner" concept that has a heritage going back to the AEC's 1972 Project LINUS. The theory behind it is complicated but appears sound. GF hopes to show it will all work in the mid-2020s.

Figure 6.5. In General Fusion's design, pistons drive a liquid wall inward to implode an FRC. *Image courtesy of General Fusion.*

6. Lockheed Martin. In 2010, under the inspiration of Dr. Tom McGuire and Charles Chase, Lockheed Martin initiated its own "Compact Fusion Reactor" development program using internal funds. The CFR appears to be a linear cylindrical system, confined at the ends by increased magnetic fields (or "magnetic mirrors") but with an extra pair of superconducting magnetic coils operating inside the plasma chamber to make "cusps" that improve confinement. This creates a very attractive magnetic field configuration, but the engineering to make it work in an actual thermonuclear system seems quite challenging.

7. EMCC. In 1987, the late visionary Robert Bussard (of Bussard ramjet fame) revived a 1950s concept originated by Philo Farnsworth (the inventor of television) to use electrostatic fields, rather than magnetic fields, to confine a fusion plasma. The idea works well enough that a very simple system can be used to generate a lot of fusion reactions, as demonstrated by neutron production, but all sorts of bells and whistles, including auxiliary magnetic fields, are needed to get it close to generating net power. Bussard managed to get preliminary funding from the US Navy, but now he is gone, and the rest of the team, led by Dr. Paul Sieck and Dr. Jaeyoung Park, are seeking private funding. Any takers?

8. Others. In addition to the above, there are quite a few dark horses in the race. These include New Jersey–based Lawrenceville Plasma Physics Fusion, led by Dr. Eric Lerner, who has produced interesting results using a concept called the plasma focus; CT Fusion, a University of Washington–based project founded by Dr. Tom Jarboe, Dr. Aaron Hossack, and Derek Sutherland, which is pursuing an FRC-like approach known as a "spheromak"; Applied Fusion Systems, founded in 2015 by Richard Dinan and Dr. James Lambert, who are trying their luck with an ST; and Hyper V, Numerex, and the Sandia Lab/ University of Rochester–based MagLIF project, which are all attempting to develop variants of the imploding liner concept.

To reach the stars, we will need what drives the stars. Thanks to the space launch revolution, we may soon have it.

Chapter 7

REACHING FOR THE STARS: WORLDS WITHOUT LIMITS

INTRODUCTION

Interstellar travel is the holy grail of astronautical engineering. The challenge is daunting, but the rewards are potentially infinite.

The most obvious challenge is that of distance. Distances to the nearest known stars are tens of thousands of times greater than those to the furthest planets in our solar system. The Earth travels at a distance of 150 million kilometers, or one astronomical unit (AU), from the sun. Mars orbits at 1.52 AU, Jupiter at 5.2, Saturn at 9.5, Uranus at 19, Neptune at 30, and Pluto at 39.5. In contrast, the nearest known stellar system, Alpha Centauri (consisting of the sunlike type G star Alpha Centauri A and the dwarf stars Alpha Centauri B and Proxima Centauri) is 4.3 light-years, or 270,000 AU, distant. Our fastest spacecraft to date, Voyager, took thirteen years to reach Neptune. That's an average of about 2.5 AU per year. However, since it employed successive gravity assists to speed up at Jupiter, Saturn, Uranus, and Neptune, Voyager managed to depart the solar system with a final velocity of about 3.4 AU per year (17 km/s). At that rate, it would take more than 790 centuries to reach Alpha Centauri. If such a probe had been launched from Earth the day *Homo sapiens* first set foot in Europe, it would still have another thirty thousand years to go.

Furthermore, not only is it far between stars, but it is hard to find much in the way of supporting resources along the way. Sunlight in deep space is nil, so power for an interstellar spacecraft must be nuclear. Even there, the answer is not easy—a typical space nuclear reactor can power itself at 100 percent capacity for only seven years or so on a full load of fuel. That's great compared to the few hours or days possible from any system burning chemical fuels, but insignificant should the output requirement be for tens of millennia.

Even communication is difficult. For example, NASA's state-of-the-art Mars Reconnaissance Orbiter uses one hundred watts of power and a two-meter-diameter dish to transmit data via X-band radio at a rate of six megabytes per second from Mars to one of the seventy-meter-diameter Deep Space Network receiving stations on Earth. If the same gear were to be used to transmit data at the same rate from Tau Ceti, some ten light-years away, the power needed would be one hundred trillion watts (a hundred terawatts!), or roughly four times all the power currently employed by human civilization.

In the face of such imposing challenges, a literature has been created showing interstellar travel as dependent upon the exploitation of exotic or fantastical physical phenomena such as wormholes, space warps, cosmic strings, and so forth. While some of these concepts are mathematically consistent with the currently known laws of the universe, there is no evidence that they exist, or if they do, that there is any method by which they could be manipulated by humans to produce a practical technology for space propulsion.

Therefore, many people believe that interstellar travel is impossible.

I disagree. Interstellar travel is incredibly difficult; perhaps as difficult to us today as a flight to Mars would have appeared to Christopher Columbus or other transoceanic navigators five hundred years ago. Indeed, the ratio of distance from Earth to Mars compared to Columbus's voyage from Spain to the Caribbean—eighty thousand to one—is roughly the same as the ratio of the distance to Alpha Centauri compared to a trip to Mars. Thus, the key missions required to establish humanity successively as a Type I, Type II, and Type III civilization all stand in similar relation to each other, and if the five hundred years since Columbus have sufficed to multiply human capabilities to the point where we now can reach for Mars, so a similar span into the future might be expected to prepare us for the leap to the stars. It should not take so long, because with its much larger population of inventive minds and better means of communication, the Type II civilization that will spread throughout our solar system over the next several centuries should be able to generate technological progress at a considerably faster rate than was possible for the emerging Type I civilization of our recent past.

I'm all for breakthroughs in physics that will give us capabilities yet unknown. We may well get them someday. But even without such, methods can already be seen in outline by which currently known physics and greatly developed and refined versions of currently understood engineering can get us to the stars. That development and refinement will occur as part and parcel of the process of the maturation of humanity as a Type II species.

When mature, Type II civilizations give birth to Type III civilizations. Here's how we'll do it.

CHEMICAL PROPULSION

Since we currently have efficient chemical rocket systems, it is worth asking if these can be used to accomplish interstellar missions. On the surface, the idea seems absurd—the maximum possible exhaust velocity for a chemical rocket is about 5 km/s (our current hydrogen/oxygen rocket engines already achieve 4.5 km/s, or 90 percent of what is theoretically feasible), and as we have already discussed, the maximum practical velocity increment that can be delivered by a rocket engine to a spacecraft is about twice the exhaust velocity. So, an advanced chemical rocket system might be able to give us a 10 km/s push, which is 2 AU per year, or 135,000 years to reach Alpha Centauri.

This is a bit on the slow side, but as we have seen, by using planetary gravity assists, Voyager was able to leave the solar system with about double this speed, reducing the required flight time to a mere seventy-nine thousand years. Obviously, this is still unacceptable, but Voyager wasn't trying for a high solar system escape speed, and all its gravity assists were done without active thrusting.

If we wanted to push the limit, we could send a spacecraft to Jupiter and use a gravity assist there to send a well-insulated, thermally protected spacecraft on a screaming dive into the inner solar system on a path that would take it just forty thousand kilometers above the surface of the sun. At that altitude, the escape velocity would be 600 km/s, and since the spacecraft would have fallen there from Jupiter, it would be traveling almost that fast at closest approach. That's when we hit the jets.

The result: after firing our engine to create a ΔV of 10 km/s at lowest altitude and then slowing down during the climb away from the sun, we will still have a departure velocity of 110 km/s, nearly seven times that of Voyager, allowing us to reach Alpha Centauri in 12,300 years.

If we really forced the engineering to wild extremes and piled on numerous stages, we might, in principle, be able to generate a ΔV of 25 km/s. This would result in a departure velocity of 175 km/s, or 7,700 years to Alpha Centauri. With chemical rocket technology, that's as good as it gets.

There are two methods that have been proposed to enable space voyages of this length. One is to put the crew in suspended animation, perhaps cryogenically frozen so they do not age. There are massive problems using this latter technique, because water expands when it freezes, thereby causing cell walls to rupture when a body is frozen. Using drugs to induce hibernation is probably possible, as illustrated by woodchucks, but aging and metabolism would still proceed, albeit at a reduced rate, making such expedients of marginal value for millennia-long voyages.

The other method is to build a spaceship large enough to house a sizable number of people for their entire lives—perhaps a nuclear-powered O'Neill colony—and send it on its way. The initial crew would raise a generation of children to carry on, who would raise another, and so on for 7,700 years until the destination star is reached. While the engineering of such a vessel would be formidable, there is nothing in the laws of physics or biology that would preclude such a mission. However, the idea that the sense of purpose of the initial crew could be preserved generation after generation for a span greater than that of all recorded human history seems rather fantastical. We therefore turn our attention to more advanced propulsion concepts that can reduce the travel time to the stars to no more than one or two human lifetimes.

FISSION PROPULSION

The fundamental physical reason why chemical rocket engines cannot produce exhaust velocities greater than 5 km/s is because the energy per unit mass, or enthalpy, of chemical fuels is limited to about thirteen megajoules per kilogram (MJ/kg) by the laws of chemistry. Nuclear fission, on the other hand, offers fuels with an energy per unit mass of 82 million MJ/kg, more than six million times as great as the best possible chemical propellants. Now, the maximum theoretical exhaust velocity of a rocket propellant is equal to the square root of twice the enthalpy; thus 5.1 km/s for chemicals, 12,800 km/s for nuclear fission. That's a lot better. The speed of light is 300,000 km/s. A fission rocket could thus, in principle, generate an exhaust velocity of 4 percent the speed of light. Since a spacecraft can generally be designed to obtain a ΔV equal to twice its exhaust velocity, a theoretically perfect fission drive could get us to 8 percent of light speed. Since Alpha Centauri is 4.3 light-years away, that would mean a one-way transit in 54 years. If half the ΔV is used to slow down at the destination, maximum speed would be 4 percent of light, and the transit time would be increased to 108 years.

There are a number of problems, however. One of them is being able to take advantage of all the energy available. Primitive nuclear propulsion systems, such as nuclear thermal rockets, do a very poor job of this. By using a solid nuclear reactor to heat a flowing gas, the maximum exhaust velocities attained are only in the 9 km/s range—good by comparison with chemical rockets, but nowhere near the performance needed for interstellar missions. If the nuclear fuel is allowed to become gaseous (a "gas core" nuclear thermal rocket—NASA did a fair amount of work on such systems in the 1960s), exhaust velocities of 50 km/s could be achieved.[1] This would be excellent for interplanetary travel but is still not in the interstellar class. If a nuclear reactor is used to generate electric power to drive an ion engine (NEP propulsion), exhaust velocities of up to several hundred km/s could be obtained, if hydrogen is employed as propellant. But the systems required are very massive, the thrust (and thus rate of acceleration) they can produce is low, and the exhaust velocity is still not good enough for interstellar missions. What is needed is a way to turn the nuclear energy directly into thrust. One answer is so straightforward it has been known since 1945: use atomic bombs.[2]

It's clear that if you detonate a series of atomic explosives right behind a spaceship, you can push it along rather well. Of course, if you don't go about it correctly, you might also vaporize the spaceship, or blow it to pieces, or turn the crew to jelly with a hundred thousand g's of acceleration, or kill everyone on board with a lethal dose of gamma rays. As we say in the engineering business, "These concerns need to be addressed." So you must do it correctly. But if you can, you've got yourself one hell of a propulsion system.

This was the idea behind Project Orion, a top secret program funded by the US Atomic Energy Commission that ran between 1957 and 1963. The original idea came from Los Alamos bomb designer Stanislaw Ulam, and the program drew the talents of such visionary weapon makers as Ted Taylor and Freeman Dyson. A diagram of one of the Orion designs is shown in figure 7.1. In it, a magazine filled with nuclear bombs is amidships. A series of bombs is fired aft down a long tube to emerge behind the "pusher plate," a very sturdy object backed up by some heavy-duty shock absorbers. When the bomb goes off, the pusher plate shields the ship from the radiation and heat and takes the impact of the blast, which is then cushioned by the shock absorbers. Since the bombs are detonated one after another in rapid succession, the net effect would be that a fairly even force is felt by the ship and its payload and crew, who are positioned forward of the bomb magazine. The pusher plate scheme is much less efficient at converting explosive force to thrust than a conventional bell-shaped rocket nozzle (perhaps only 25 percent, compared to the 94 percent that is state-of-the-art), but it has much more force to play with. So maybe the real effective exhaust velocity would only be about 1 percent the speed of light. That puts a bit of a crimp on our plans for fission-driven interstellar flight, but still, an exhaust velocity of 3,000 km/s in a high-thrust rocket has got to be considered pretty good.

However, for better or for worse, the Orion project came to a screeching halt in 1963 when the Limited Nuclear Test Ban Treaty between the United States and the Soviet Union banned the stationing or detonation of nuclear weapons in outer space.

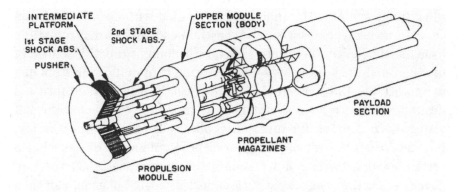

Figure 7.1. Orion nuclear-bomb-driven spacecraft. The bombs are dropped down a center tube and explode behind the pusher plate, which absorbs the shock and shields the payload section. *Image courtesy of the US Atomic Energy Commission.*

The treaty will expire someday. But still, it seems like a good idea to avoid stationing ships in space filled with thousands of atomic bombs (and an even better idea to avoid having factories mass-producing such bombs for sale to space travelers). I proposed a way around this problem in the early 1990s with a concept called a nuclear salt-water rocket (NSWR), shown in figure 7.2.[3]

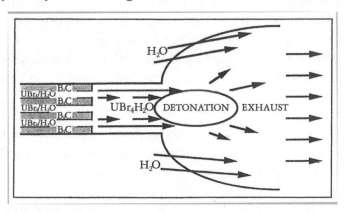

Figure 7.2. Nuclear salt-water rocket.

In the NSWR, the fissionable material is dissolved in water as a salt, such as uranium bromide. This is stored in a bundle of tubes, separated from each other by solid material loaded with boron, which is a very strong neutron absorber and therefore cuts off any neutron traffic from one tube to another. Since each tube contains a subcritical mass of uranium, and the boron cuts off any neutron communication from one to another, the entire assembly is subcritical. However, when thrust is desired, valves are open simultaneously on all the tubes, and the salt water, which is under pressure, shoots out of all of them into a common plenum. When the moving column of uranium salt water reaches a certain length in the common plenum, a "prompt critical" chain reaction develops, and the water explodes into nuclear heated plasma. This then expands out a rocket nozzle that is shielded from the heat of the plasma flow by a magnetic field. In effect, a standing detonation similar to chemical combustion in a rocket chamber is set up, except that the enthalpy available is millions of times greater. The nozzle would be much more efficient than the Orion pusher plate, but because the uranium content of the propellant is "watered down," the exhaust velocity would also be decreased significantly below nuclear fission's theoretical maximum of 4 percent of light speed, perhaps to about the same 1 percent achievable by an Orion driven by a nuclear fission bomb. But at least the need for mass-produced bombs would be eliminated.

With exhaust velocities of about 1 percent the speed of light, interstellar spaceships driven by such systems might be able to attain 2 percent light speed, allowing Alpha Centauri to be reached in about 215 years. Voyages with trip times on this order might be able to use rotating hibernations to allow a crew to reach the destination. Alternatively, there at least would be some chance that a multigeneration interstellar spaceship could reach its goal with its purpose remaining intact.

However, in addition to offering only marginal performance for interstellar travel, such fission drives have another problem—fuel availability. The amount of fissionable uranium-235 or plutonium-238 needed to fuel such systems would be enormous—perhaps ten thousand tons to send a one-thousand-ton (small for a slow, long-duration interstellar spaceship) payload on its way. It is unclear where such supplies could be obtained. One possibility might be to breed fissionable U233 from Th232 using spare neutrons from either a thorium-fueled fission reactor (which are currently being actively pursued by entrepreneurial fission power start-ups) or a D-D fusion reactor. But still, it would be a very large logistic requirement.

We therefore turn our attention to a still more potent source of energy for interstellar spaceship propulsion, thermonuclear fusion.

FUSION PROPULSION

High exhaust velocity is key to interstellar rocketry, and enthalpy is the key to exhaust velocity. Nuclear fission looks attractive at 82 million MJ/kg, but nuclear fusion is better. For example, if pure deuterium is used as fuel and burned together with all intermediate fusion products (a series of reactions known as "catalyzed D-D fusion"), 208 million MJ/kg of useful enthalpy is available for propulsion, plus 139 MJ/kg of energetic neutrons, which, while useless for propulsion, can be used to produce onboard power. If a mixture of deuterium and helium-3 is used as fuel, the useful propellant enthalpy is a whopping 347 million MJ/kg. As a result, thermonuclear fusion using catalyzed D-D reactions has a maximum theoretical exhaust velocity of 20,400 km/s (6.8 percent of light speed, or 0.068 c) while a rocket using the D-He3 reaction could theoretically produce an exhaust velocity of 26,400 km/s, or 0.088 c.

Now we're talking starflight! With quadruple the enthalpy of nuclear fission and much more plentiful fuel, nuclear fusion holds the potential for a real interstellar spaceship propulsion system. As in the case of nuclear fission, fusion offers both pulsed explosions and steady-burn options for rocket propulsion, but in the case of fusion, both are more practical to implement.[4]

Fission bombs must be of a certain minimum size, because for a fission chain reaction to occur, a "critical mass" of fissile material must be assembled. Unless one chooses to simply waste energy by designing an inefficient explosive (a choice that is not a viable option for interstellar propulsion), this critical mass implies a minimum yield for a fission bomb of about a thousand tons of dynamite.

Fusion is different. There is no critical mass for nuclear fusion, so in principle, fusion explosives could be made as small as desirable. Current military fusion explosives—H-bombs—have very high yields, because they use a fission atomic bomb to suddenly compress and heat a large amount of fusion fuel to thermonuclear detonation conditions. If one wished to be crude, one could use such hydrogen bombs in an Orion-type propulsion system, with considerably higher performance and much cheaper fuel than the A-bomb version. However, with fusion there are other ways to achieve the required detonation effect on a much smaller scale.

For example, one can use a set of high-powered lasers to focus in on a very small pellet of fusion fuel, thereby heating, compressing, and detonating it. Preliminary experiments have proven the feasibility of such "laser fusion" systems, and one, the National Ignition Facility or NIF, at the Livermore Lab in California, has demonstrated about one-third the power needed to ignite thermonuclear explosions. An interstellar spaceship utilizing such a system for propulsion would eject a series of pellets with machine-gun rapidity into an aft region of diverging magnetic field. As each pellet entered the target zone, it would be zapped from all sides by an array of lasers. It would then detonate with the force of a few tons of dynamite, and the ultrahot plasma produced would be directed away from the ship by a magnetic nozzle to produce rocket thrust.

(See plate 11.)

Alternatively, it may be possible to implode and detonate fusion pellets using an appropriately shaped set of chemical explosives. I say "may" because a great deal of top secret work has been done to achieve this goal in both the United States and the former Soviet Union, but the results are unpublished. If feasible, such chemically ignited fusion micro-bomblets would eliminate the need for a heavy laser system aboard ship.

As a third alternative, one could implement fusion propulsion without bombs, lasers, or micro-bomblets by using a large magnetic confinement chamber to contain a large volume of reacting thermonuclear fusion plasma. This is presumably the type of system that would be used to produce fusion power in the future, except in such a fusion drive, most of the ultrahot (tens of billions of degrees, or several megavolts) fusion products would be allowed to leak out of one end of the reactor to produce thrust, while the rest would be used to heat the plasma to 500,000,000°C (fifty kilovolts) or so, which is the proper temperature for fusion reactors. Some of the lower-temperature plasma would also leak out, but because of its lower energy, it could be decelerated by an electrostatic grid and used to produce electric power for the ship.

(See plate 12.)

The magnetic nozzles used by fusion propulsion systems would not be as good as the 94 percent efficient bell nozzles used in chemical rocket engines but would be much more effective at channeling thrust than the 25 percent efficient pusher plates of the old Orion. Probably an efficiency of about 60 percent could be achieved. Assuming that to be the case, then a D-He3 fusion rocket should be able to attain an exhaust velocity of about 5 percent the speed of light. Since practical spacecraft can be designed to reach a speed about twice their engines' exhaust velocity, this implies that such fusion propulsion systems could make 10 percent of light speed. Ignoring the small amount of extra time needed to accelerate, that means one-way to Alpha Centauri in forty-three years—or eighty-six years, if we need to use the propulsion system to slow down.

ANTIMATTER

While deuterium-helium-3 fuel has the highest enthalpy of any substance that can be found in nature, there is an artificial material that has a much higher enthalpy still: antimatter.[5]

Antimatter is mass with the charges of the subatomic particles reversed. In ordinary matter, electrons are negative; in antimatter, they are positive. Ordinary protons are positive; antiprotons are negative. Because oppositely charged particles attract, antiparticles attract their ordinary-matter mates. The attraction is fatal, though, as the two annihilate one another, transforming their combined mass into energy in accord with Einstein's famous formula $E = mc^2$ (energy equals mass times the speed of light squared).

Antimatter is such a staple of science fiction that many people believe that it *is* science fiction, but antimatter is real. We don't ordinarily encounter it in daily life because the universe, or at least our region of it, was created with an excess of ordinary matter over antimatter—all the antimatter (or all the antimatter in our galaxy) has been annihilated, leaving nothing but the common stuff. But because energy can also be turned into matter in accord with the Einstein formula, evanescent antiparticles are created by cosmic ray impacts with the Earth's atmosphere. We have also been able to create antiparticles in high-energy accelerators and have succeeded in combining antiprotons with antielectrons (or positrons) to produce antihydrogen atoms. These antihydrogen atoms have been further combined to form antihydrogen molecules. Antiprotons can be stored in special jars called "Penning traps," in which magnetic fields are used to keep the ions from hitting the wall (where they would annihilate). In this way, up to several million antiprotons at a time can be stored for extended periods. Using the big collection rings at leading high energy physics accelerator facilities such as Fermilab and CERN, up to a trillion antiprotons at a time have been collected. This represents about 1.7 picograms (a picogram is a trillionth of a gram) of antimatter. If this much antimatter were allowed to annihilate, it would release about three hundred joules of energy, enough to light a sixty-watt light bulb for five seconds. Tiny amounts of antihydrogen atoms and molecules have also been confined, using the pushing power of lasers to herd them away from chamber walls.

Now, let's say we could do much better than this and freeze antihydrogen gas into solid crystals. We could then give these crystals a static electric charge, allowing us to store them without touching by levitation inside a magnetic or electrostatic trap. Then we could use this material as fuel on an interstellar spaceship, annihilating it with ordinary hydrogen to produce energy. How much energy? Lots. Because the speed of light, c, is such a large number—300,000 km/s— Einstein's formula is generous. If we were to annihilate a single half kilogram of antimatter with a half kilogram of ordinary matter, we would release ninety billion megajoules of energy. That's an enthalpy of 90 billion MJ/kg, 259 times greater than D-He3 fusion, more than a thousand times greater than nuclear fission, and nearly seven billion times as great as an equivalent amount of hydrogen/oxygen rocket propellant. Put another way, a single kilogram of antimatter annihilating with a kilogram of ordinary matter will release as much energy as forty million tons of TNT. The maximum theoretical exhaust velocity of an antimatter rocket would be the speed of light.

That's theory; in practice, things are not quite that good. In the first place, about 40 percent of the energy from antimatter annihilation is released in the form of gamma rays with energies of more than two hundred million electron volts. This is hundreds of times greater than the typical gamma rays released by nuclear fission reactors and will put a very heavy shielding burden on the spacecraft. Then there is the issue of how thrust will be created. One idea would be to use antimatter to generate an extremely high-energy magnetically confined plasma, with average energies of hundreds of millions of electron volts. In this case, only the portion of the antimatter annihilation energy that comes off as charged particles would be usable, since the gamma rays and uncharged particles would escape from the system before they could heat the plasma. In addition, such a high-temperature plasma would waste massive amounts of energy through both cyclotron and bremsstrahlung radiation. These losses, taken together with the roughly 60 percent efficiency possible with magnetic nozzles, would reduce the attainable effective exhaust velocity of an antimatter plasma drive down to perhaps 30 percent the speed of light.

An alternative method of antimatter propulsion would be to use the energy of annihilation to heat the surface of a stern-mounted solid cylinder composed of a high-temperature material such as graphite or tungsten to incandescence, and then direct the light radiated by the glowing object rearward with mirrors. Alternatively, matter and antimatter could be injected to collide at the focus of a parabolic mirror, which would direct the light backward (see figure 7.3). Particles of light, called photons, have momentum, and if they are all directed rearward, a net forward force would be created. Such a system is termed a photon rocket.

The exhaust of a photon rocket has the speed of light (because it is light), but not all the energy of the antimatter annihilation will go into it. Most of the energy of the highly penetrating gamma rays, neutrinos, and other highly penetrating uncharged particles will carry their energy out of the system before it can do any good. This reduces the effective exhaust velocity (specific impulse) of the system considerably. But antimatter has energy to spare. Even when all the losses are taken into account, effective exhaust velocities for photon rockets on the order of 50 percent light speed appear attainable.

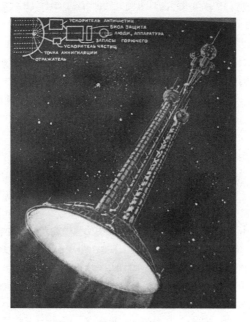

Figure 7.3. Soviet photon rocket concept.

The photon rocket is simpler than an antimatter plasma drive and offers higher performance. Therefore, if antimatter does become available in sufficient quantities to power interstellar voyages, photon rockets will probably be the engines of choice.

But availability is an issue. Using our current accelerator-based techniques for manufacturing antimatter, it requires more than ten million times as much electric power to create a unit of antimatter as the antimatter energy is worth.

Consider what this means. Let's say we want to get a one-thousand-ton dry mass interstellar spaceship up to 10 percent the speed of light. Assuming we use a fusion rocket with an exhaust velocity 7 percent the speed of light to get it up to speed over a ten-year period, we will need an average acceleration of 0.01 g/s, a thrust of about two hundred thousand newtons (because it also has about three thousand tons of propellant on board at the start of the mission), and a jet power of about two terawatts. The total amount of energy expended would be about twenty terawatt-years—or about what the entire human race consumes in a year right now. At current electric power prices of $0.06/kilowatt-hour, this much energy would be worth about $10.5 trillion, roughly the current US government national budget for three years. The fusion fuel required to produce the power might have a value of 10 percent of this, or $1 trillion, but this figure could be doubled due to the inefficiency of the drive. So, bottom line, say $2 trillion for the mission propellant price using fusion fuels. This is rather costly, but a rich, well-developed, solar-system-spanning Type II civilization could be able to afford it for a project as important as colonizing another stellar system.

However, if current accelerator-based systems were used to produce antimatter fuel for this mission, the energy costs for the efforts would be multiplied ten million times over. There are techniques under discussion within the antimatter community that could increase the efficiency of production significantly, perhaps as much as a factor of a thousand. But even with this improvement, the cost of antimatter propellant would be ten thousand times that of the same mission using fusion fuel.

Of course, if one wanted to go significantly faster than 10 percent the speed of light, fusion drops out of the picture because the maximum exhaust velocity of about 0.07 c is insufficient. As discussed above, the effective exhaust velocity of an antimatter photon rocket would be about 50 percent the speed of light, making flights at up to 90 percent of light speed theoretically possible. This would get a ship to Alpha Centauri in about five years, which would seem like three to the crew due to the effects of relativistic time dilation. But the society that launched such a mission would have to be one that was so rich that cost was simply not an issue.

LIGHT SAILS

Nearly four hundred years ago, the famous German astronomer Johannes Kepler observed that regardless of whether a comet is moving toward or away from the sun, its tail always points away from the sun. This caused him to guess that light emanating from the sun exerts a force that pushes the comet's tail away. He was right, although the fact that light exerts force had to wait till 1901 to be proven by Russian physicist Peter N. Lebedev, who made mirrors suspended on thin fibers in vacuum jars turn by shining light upon them. A few years later, Albert Einstein provided the theoretical basis for this phenomenon, explaining why light exerts force in his classic paper on the photoelectric effect, for which he later received the Nobel Prize.

Well, if light can push comet tails around, why can't we use it to move spaceships around? Why can't we just deploy big mirrors on our spacecraft—solar sails, if you will—and have sunlight push on them to create propulsive force? The answer is that we can, but it takes an awful lot of sunlight to exert any significant amount of push. For example, at 1 AU, the Earth's distance from the sun, a square solar sail one kilometer on a side would receive a total force of 10 newtons, about 2.2 pounds, pushing on it from the sun. For such a large object, that's not a lot of force. Consider, if the one-square-kilometer sail were made of plastic as thin as writing paper (about 0.1 mm), it would weigh one hundred tons, and it would take a full year for sunlight at 1 AU to accelerate it through a ΔV of 3.2 km/s.

This is not an especially impressive performance, but writing paper is hardly the thinnest thing we can manufacture. Let's say we made the sail 0.01 mm thick (ten microns—depending upon the brand, kitchen trash bags are twenty to forty microns thick). This is about the thickness of the films used on many high-altitude balloons. In that case, the sail would only weigh ten tons, and it could accelerate itself 32 km/s—roughly the round-trip ΔV needed to go from low Earth orbit to Mars and back on a low-thrust trajectory, in just about a year. Of course, if the sail were hauling a payload equal to its own weight, that would slow it down by a factor of two. Still, a ten-micron-thick solar sail would be in the ballpark for an effective propulsion device supporting Earth-Mars transportation.

The advantage of the solar sail is clear—it needs no propellant or onboard power supply. Fundamentally, the technology is simple, cheap, scalable, elegant, and, in a word, beautiful. The idea of ships equipped with huge ultralightweight shiny sails coursing effortlessly through space on the power of reflected sunlight alone is romantic in the extreme, recalling as it does the age when sailing ships opened the oceans of Earth to explorers, merchants, and adventurers of every type. Moreover, solar sails may hold enormous potential to similarly open the lanes of interplanetary commerce. For this reason, many people, including noted science fiction author and space visionary Arthur Clarke and former Planetary Society executive director Louis Friedman, have long been staunch advocates of this technology.[6] Its development for interplanetary propulsion purposes does not appear to be especially formidable, being mostly a matter of mastering some packaging and mechanical deployment issues, and the fact that solar sails are not yet in general use is abundant testimony to the stagnation in the space program over the past several decades. That said, in 2010, the Japan Aerospace Exploration Agency actually flew a small solar sail spacecraft called IKAROS from Earth to Venus, demonstrating the technology in interplanetary flight for the first time. Certainly, a mature Type II civilization will not only possess solar sails but employ them widely for interplanetary commerce and numerous other applications.

(See plate 13.)

But we are talking about interstellar propulsion here. How can a system that derives its motive force from the sun be used to drive a ship through the darkness of interstellar space?

One answer, the simplest and most elegant, is simply to make the solar sail so thin that it can use sunlight to accelerate the spacecraft to interstellar speeds while it is still within the solar system. Such ultrathin solar sails would have to be manufactured in space using techniques that are currently unavailable. To save weight, we would discard the plastic backing and just use a thin layer of aluminum created by molecular deposition in vacuum on a lightweight webbing for the sail. The following table shows the maximum speed that such a system could achieve driving a thousand-ton spacecraft, assuming that the payload spacecraft has a mass equal to that of the sail and that the mission begins 0.1 AU from the sun.

TABLE 7.1. THIN SOLAR SAILS FOR INTERSTELLAR TRAVEL

Sail thickness	Acceleration at 1 AU	Sail radius	Final velocity
0.3 microns	0.006 m/s²	220 km	95 km/s (0.03 percent c)
0.1 microns	0.018 m/s²	234 km	212 km/s (0.07 percent c)
0.01 microns	0.18 m/s²	2,108 km	728 km/s (0.26 percent c)
0.001 microns	1.8 m/s²	2,343 km	2,322 km/s (0.77 percent c)

It's hard to get the solar sail much thinner than 0.001 microns, because this thickness represents a layer of material just four atoms across. (In fact, to avoid being transparent, the aluminum probably must be at least 0.01 microns thick, but an average density equivalent to a sail 0.001 microns thick can nevertheless be achieved by perforating the sail. Provided the holes are a lot smaller than the 0.5-micron wavelength of visible light, it will still reflect light in the same way that a chicken-wire radio antenna reflects radio waves.) Starting the mission closer to the sun than 0.1 AU is conceivable, but the final speed will only increase in proportion to the inverse square root of the distance (i.e., if we get nine times closer, we will only end up going three times faster), and the acceleration for our 0.001-micron spacecraft when it starts at 0.1 AU is already a stiff 18 g's. So the bottom line is that a light sail driven only by sunlight is unlikely to be able to get an interstellar spaceship much above 1 percent the speed of light. An advantage of such a system would be that it would be cheap (energy cost is zero), simple, and reliable, and provided the target star has comparable luminosity to the sun, the same solar sail used to accelerate the mission could also be used to decelerate at the destination star, navigate within its solar system, or even to return. But the flight time to Alpha Centauri would be on the order of five centuries. Perhaps such a system might be acceptable for a multigeneration ship or one employing suspended animation techniques. Perhaps. It might also be found acceptable for use by a species whose natural life span is considerably longer than current humans'.

But if they are to serve as a practical means of interstellar propulsion for people as we know them, light sails will need an additional shove to get to speed. One way to do that would be to push them with high-energy lasers, an idea first proposed by physicist Robert Forward in 1962.[7]

Let's take the thousand-ton interstellar spaceship together with the 343-kilometer-radius, 0.001-micron-thick light sail discussed above, and illuminate the sail with laser light five times as bright as sunlight is on Earth (i.e., about as bright as sunlight is at 0.45 AU). The ship will then be accelerated at the comfortable clip of 9 m/s^2 (0.92 g's), reaching 15 percent the speed of light inside of two months. At the end of that time, the ship will have traveled 121 billion kilometers, or 806 AU. In order to keep focused on the light sail at this distance, the laser projector would have to have a lens about one hundred meters in radius. This is only about twelve times larger than the largest telescope yet built or under construction (the Keck sixteen-meter diameter) and so may be considered a modest challenge compared to the rest of the project. However, the amount of power the laser would need is formidable: 240 terawatts. This is about ten times the total power humanity today generates. However, since it would only be needed for two months (i.e., one-sixth of a year), the total energy would be about what humanity currently consumes in two years. Obviously, even if we had the technology, such an expenditure of power would be out of the question today. But humanity's power production is growing at a rate of 2.6 percent per year. If this trend continues, in the year 2200 we will be producing and consuming energy at a rate of 2,500 terawatts, and using 240 TW, or 9.6 percent, of this for two months to get to the stars might well be considered affordable.

An alternative approach to making laser light sail propulsion affordable would be to lighten the spacecraft. This is the approach that has been adopted by the Breakthrough Starshot mission program, initiated in 2015 with funding from Russian billionaire Yuri Milner. Rather than try to push a thousand-ton crewed interstellar spaceship, the Breakthrough Starshot mission proposes to make use of advanced microelectronics to create an unmanned probe with a mass of just one hundred grams, sail included. With the mass cut by a factor of ten million, the power requirements would be as well. So instead of needing a 240-terawatt laser, a 24-megawatt unit would do the job. Such a combination of microspacecraft, laser, and sail technology could well become available within the next decade or two, putting this exciting precursor mission for interstellar exploration on the agenda for our time.

The laser projector would be kept pointed at the target star. The crew or computer aboard the spacecraft would know in advance the position of the projector as it orbits the sun and use this knowledge to keep their vessel squarely in the center of the light beam. At 15 percent the speed of light, they would reach Alpha Centauri in about twenty-nine years. If the laser lens were four times as big (i.e., four hundred meters instead of one hundred meters), we could keep the light on the sail for twice as long, go twice as fast (30 percent c!), and reach the destination in half this time.

There's just one problem: no way to stop.

SATURN EXPRESS

It seems to me that the Breakthrough Initiative needs a near-term mission that can act as a transition to its visionary goal of interstellar travel. I have a concept for this, which I call "Saturn Express." The basic idea is to create an ultrafast sail craft that can enable near-term exploration of the outer solar system while demonstrating the potential for more advanced incarnations to go to 550 AU, and then the stars.

So consider: at 1 AU, solar light pressure is nine micronewtons per square meter. So a sail craft with an areal density of 1.5 g/m² would experience an acceleration away from the sun of 0.006 m/s², which is exactly the same as the gravitational acceleration exerted by the sun at 1 AU. These two forces would balance, at 1 AU and all other distances, since they both go down with the inverse square of the distance. Therefore, such a craft released with escape velocity from Earth would move out in a straight line along the tangent to Earth's orbit at the Earth's orbital speed, which is 30 km/s, or 6 AU/yr. It would thus be able to reach Jupiter in about 0.8 years, Saturn in 1.5 years, Uranus in 3.2 years, and Neptune in 5 years.

Saturn would be an excellent first target because it is of such great interest to exobiology. If we have a sail with an areal density of 1 g/m^2 (one micron thick, which should be doable) and an area of one hundred square meters, it will have a mass of one hundred grams. This leaves fifty grams for the spacecraft. The configuration could be like a parachute, with the sail craft like the parachutist astern of the sail, which would be billowing out ahead of it. This would be passively stable, with the concave side of the parachute pointing toward the sun, which would also be toward the Earth once it was out far enough, providing a high gain reflector. The sail craft would mount a strobe light, blinking with a power of one hundred watts for one millisecond every one thousand seconds. It would have an average power consumption of 0.0001 watts, using up 0.88 watt-hours per year. By looking at it with Hubble or Webb, we could track its speed via its Doppler shift. If the craft could be targeted to fly behind Enceladus, we conceivably might be able to look for the spectrum of organic molecules launched into orbit by its plumes.

That's the basic idea. By leaving the multimegawatt laser for later, Breakthrough Starshot could have something flying soon, on a budget well within the means of Yuri Milner. It would be a profound demonstration of fast sail technology, directly traceable to more advanced versions with potential for interstellar missions.

I sent a write-up laying out this plan to the Breakthrough Starshot Foundation in 2018. Let's hope they go for it.

MAGNETIC SAILS

In 1960, the visionary physicist Robert Bussard published one of the classic papers on interstellar travel.[8] In it he proposed a kind of fusion ramjet that would gather interstellar hydrogen as it flew, then burn it to produce thrust using the same proton-proton fusion reaction that powers the sun. Bussard's concept was elegant. It was based on more-or-less known physics, yet because the propellant was gathered in flight, there was no mass ratio limit, and the spacecraft could accelerate continuously to asymptotically approach the speed of light.

There are a number of problems, however. One is that the proton-proton reaction is very hard to drive and occurs slowly, so that igniting a fusion reactor using this fuel is twenty orders of magnitude more difficult that the deuterium-tritium, deuterium-deuterium, or deuterium-helium-3 systems that humanity is currently struggling to make work. Astronomer Daniel Whitmire in 1975 proposed an improvement by recommending that carbon be added to catalyze the reaction using the same carbon-nitrogen-oxygen (CNO) catalytic fusion cycle that drives the process of proton fusion in certain hot stars. This raises the reactivity of the system to the point where it is only a million times more difficult to ignite that deuterium, which certainly helps, but even with CNO catalysis, artificial proton fusion reactors remain a difficult and distant prospect.

The other problem is how to gather the material. Because of the diffuse nature of the interstellar medium, the scoop has to be huge, so using a physical inlet is out of the question. The only viable options seem to be some sort of scooping device based on magnetic or electrostatic fields.

In 1988, Boeing engineer Dana Andrews decided to try to take a small step toward a Bussard ramjet by proposing a concept in which a magnetic scoop would be used to gather hydrogen ions in interplanetary space for use as propellant in an ordinary ion engine, which would be powered by an onboard nuclear reactor. This concept thus eliminated the need for proton fusion required by Bussard's ramjet. The performance of the state-of-the-art nuclear electric propulsion system was too low to be relevant for interstellar missions, but for interplanetary travel, the self-fueling ion drive would be terrific. There was a problem, though. As far as Andrews could calculate, the magnetic scoop employed by the system generated more drag against the interplanetary medium than the ion engines produced thrust. The drive was apparently useless.

I was living in Seattle at the time, and Andrews and I were well acquainted. Because I have a good background in plasma physics, Andrews told me about his concept and the problem he was running into. At first I thought there might be a solution, because Andrews was using certain approximations to calculate the plasma drag that were very rough for the situation he was dealing with. So we worked together to write a computer program to calculate the drag more precisely, only to discover that the actual drag was much greater than Andrews had first estimated. At that point, I suggested that we abandon the ion thrusters entirely and, rather than seek to minimize the drag, try to maximize it— to use the magnetic field not as a scoop but as a sail. In this manner, we would derive the spacecraft's motive force from the dynamic pressure of the solar wind, the plasma that flows outward from the sun. Andrews agreed, and we went back to the drawing board with a new approach. Thus was born the magnetic sail, or magsail.[9]

The idea was timely. In 1987, Professor Ching-Wu (Paul) Chu of the University of Houston had just invented the first high-temperature superconductors, materials that can conduct electricity without any resistive power dissipation and that operate at reasonable temperatures (previously, the only known superconductors operated at temperatures approaching absolute zero). The magsail could use these to create a powerful magnetic field that could deflect the solar wind, thereby imparting its force to propel a spacecraft. If practical high-temperature superconducting wire could be developed that could conduct currents with the same density as state-of-the-art low-temperature superconductors (about a million amps per square centimeter), then magsails could be developed that could produce fifty times the thrust-to-weight ratio of near-term (10-micron) solar light sails. (This was visionary thirty years ago, but as of this writing, high-temperature superconducting wires offering current density of three million amps per square centimeter have become available, about triple that of low-temperature superconductors.) Even though magsail thrust is always nearly outward from the sun (as opposed to light sails, which can use the mirror effect to aim their thrust through a wide angle), I was able to derive equations showing how the system could be navigated almost at will throughout the solar system. The maximum possible speed of a solar-wind-pushed magsail is the speed of the solar wind—500 km/s—which is too slow for interstellar flight, and a practical magsail could probably only do half of this. But Andrews has investigated propulsion options including pushing magsails with plasma bombs ("MagOrion") and charged particle beams that offer significant promise.[10]

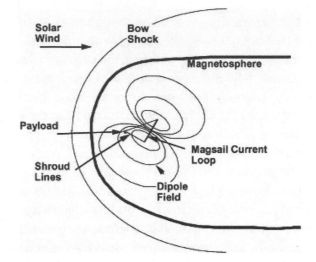

Figure 7.4. The magsail creates a miniature magnetosphere that blocks the solar wind, thereby delivering thrust to a spacecraft.

ELECTRIC SAILS AND DIPOLE DRIVES

In 2004, Finnish scientist Pekka Janhunen proposed an alternative type of plasma sail device, which he called the electric sail.[11] In this system, a grid of wires is charged up positive, causing it to deflect the protons on an ongoing plasma wind, thereby creating drag. I was one of the reviewers of his paper when he submitted it to the *Journal of Propulsion*. Despite the fact that there were many unanswered concerns raised by his original article, including the possibility that the electric sail would become self-shielded or neutralized by electrons attracted to its positive charge faster than its onboard electron gun could get rid of them, I recommended publication so the broader aerospace community could debate the issues and help find solutions. This turned out to be a good decision, because after extensive analysis by the advanced propulsion group at NASA's Marshall Space Flight Center, the concept ultimately held up, and test flights are planned for the early 2020s.

Considering this, in 2016 I came up with a further development of the electric sail concept that I called the dipole drive, which uses

two screens placed parallel to each other, one positive and one negative. In this configuration, the two screens neutralize each other's field
everywhere except in the region between them, where their fields add
together. If the positive screen is faced on the side of the sail away from
the plasma wind, the dipole drive will reflect the protons like a mirror,
propelling the ship downwind. If the positive screen is on the side of
the dipole drive facing the plasma wind, the drive will accelerate the
protons, producing thrust in the upwind direction. If the screens are
placed at an angle to the wind, a force perpendicular to the wind ("lift"
in aerodynamic parlance) can be created. Thus, unlike the magnetic
or electric sail, the dipole drive can actually push in any direction,
affording a spacecraft much more maneuverability. The dipole drive
needs power to thrust upwind but can actually be made to generate
power when being used to sail downwind.[12] Diagrams showing the
basic principle of the dipole drive are shown in figure 7.5.

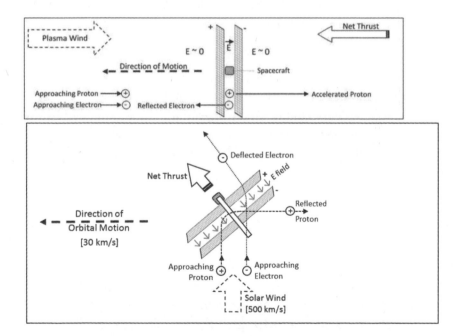

**Figure 7.5. The dipole drive can thrust upwind (*top*) or downwind
or crosswind (*bottom*) without any need for onboard propellant.
*Image courtesy of Heather Rose, Pioneer Astronautics.***

A unique feature of the dipole drive is that it is the only propellantless propulsion system that can thrust *toward* the sun. In contrast, magnetic sails, electric sails, and even solar light sails always have a component of force pushing them away from the sun. This means that a dipole drive could, at least in principle, be used to increase the sun's attraction to a spacecraft, thereby keeping it orbiting within the solar system while it accelerates to very high speeds with the help of beamed power.

However, the most interesting and important thing about these novel plasma sail systems is not their capability to speed up a spacecraft—what's important is their ability for slowing one down. A plasma sail is the ideal interstellar mission brake! No matter how fast a spaceship is going, all it has to do to stop is deploy and turn on a magsail, electric sail, or dipole drive, and drag generated against the interstellar plasma will do the rest. Just as in the case of a parachute deployed by a drag racer, the faster the ship is going, the more "wind" is felt, and the better it works.

Plasma sails thus provide the missing component needed for interstellar missions using laser-pushed light sails. Alternatively, if fusion rockets (or any other type of rocket) are used to accelerate, having a plasma sail on board means that no fuel will be needed to decelerate. All of the available ΔV can be used to speed up; none is needed to slow down. As a result, the ship can perform its mission twice as fast.

In starflight, deceleration is half the battle. Half the battle is already won.

BREAKTHROUGH PHYSICS AND
TRULY ADVANCED PROPULSION

All of the technologies I have presented thus far in this chapter are engineering applications of well-known and understood physics. As an engineer, that's what I do. I apply physics. I don't invent it.

There are, however, quite a few other concepts that have been advanced that defy known science, employ novel interpretations, or require new physical laws. In the period from 1996 through 2002,

Plate 1. February 6, 2018: Falcon Heavy takes flight, launching a new era for humanity in space. *Image courtesy of SpaceX.*

Plate 2. Hitting the road to Mars. *Image courtesy of SpaceX.*

Plate 3. Falcon Heavy boosters return to base. *Image courtesy of SpaceX.*

Plate 4. Kiwi ingenuity. Rocket Lab's Electron launch vehicle takes flight from its New Zealand launchpad for the first time on May 25, 2017. The flight reached space but failed to make orbit. But on its second launch on January 23, 2018, Electron made it all the way, successfully delivering two Spire Lemur-2 CubeSats to their desired circular orbit. Electron has a payload capacity of two hundred kilograms, making it an excellent choice to launch constellations of microsatellites. *Image courtesy of Rocket Lab.*

Plate 5. Mars Direct surface base. Habitation module is at left, energy recovery ventilator at right. *Image courtesy of Robert Murray, Mars Society.*

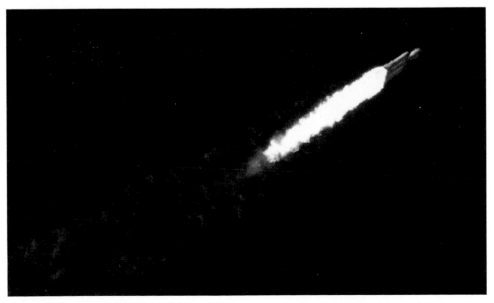

Plate 6. Once SpaceX implements its plans to refuel upper stages on orbit, the now-operational Falcon Heavy will have greater trans-Mars throw capability than a Saturn V rocket. *Image courtesy of SpaceX.*

Plate 7. The Interplanetary Transport System (ITS), SpaceX's vision for colonizing Mars. The ITS was subsequently renamed the Big Falcon Rocket or BFR, then Starship. *Image courtesy of SpaceX.*

Plate 8. In the course of things, children will be born and families will be raised on Mars, the first true colonists of a new branch of human civilization. *Image courtesy of Robert Murray, Mars Society.*

Plate 9. SpaceX illustration of its proposed 2023 artists' cruise around the moon. Provided the musicians are up to it, voyages to near-Earth asteroids could be next. *Image courtesy of SpaceX.*

Plate 10. Jupiter's moon Europa is an ocean world entirely covered by sea ice. The subsurface ocean is kept liquid by heating due to tidal interaction with Jupiter. *Image courtesy of NASA.*

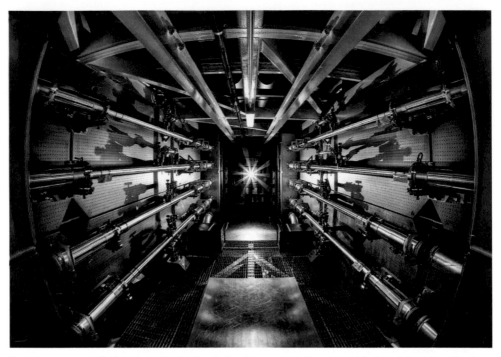

Plate 11. Laser fusion has been partially demonstrated at the National Ignition Facility. *Image courtesy of the US Department of Energy.*

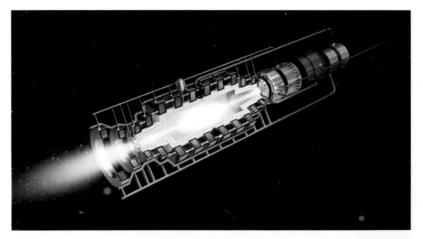

Plate 12. Magnetic Confinement Fusion Propulsion System. A magnetic field confines reacting plasma in a chamber, which is open to plasma escape at one end, thereby producing thrust. *Image courtesy of S. Shalumov, Princeton Satellite Systems.*

Plate 13. Artist's conception of the IKAROS solar sail spacecraft approaching Venus. The IKAROS sail is 7.5 microns thick and 14 meters on a side. It employs thin film solar arrays embedded in the sail for power and variable reflectance liquid crystal panels for attitude control. *Image courtesy of JAXA.*

Plate 14. The Mars of the future? Mars was once a warm and wet planet and could be made so again through human engineering efforts. But this time it would be seeded with life. We can create a living world. Would it be ethical to do it? Would it be ethical not to? *Image courtesy of Wikimedia Creative Commons, author: Daein Ballard, licensed under the GNU Free Documentation License.*

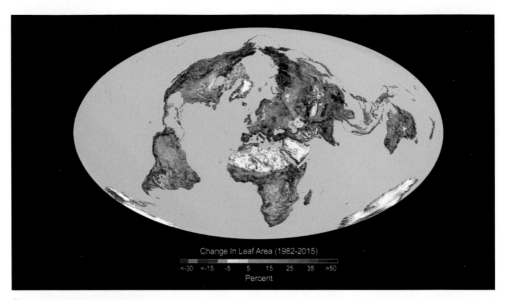

Change In Leaf Area (1982-2015)

<-30 <-15 -5 5 15 25 35 >50
Percent

Plate 15. Leaf area has increased an average of about 20 percent worldwide over the past thirty-six years, a beneficial effect of rising global CO_2 levels. But what is happening in the oceans? *Image courtesy of NASA.*

Plate 16. The Mars Society's Flashline Mars Arctic Research Station sits on the edge of the twenty-three-kilometer-diameter Haughton Crater made by the strike of an asteroid, approximately two kilometers in size, on Canada's Devon Island thirty-nine million years ago. The impact energy was two billion kilotons.

NASA set up a Breakthrough Propulsion Physics program to investigate them. None were confirmed, most were refuted, a few remain in doubt.[13] Since that time, some new ones have been energetically promoted, including the EmDrive and the Mach drive.[14] I remain quite skeptical of such claims.

That said, I do believe that there is plenty of new physics out there waiting to be revealed that could have tremendous consequences for propulsion. As I discuss in chapter 9, the existence of the universe, its laws, its fundamental particles, and all the matter and energy in it remains unexplained. Titanic forces and energies may—indeed, almost surely *must*—exist of which we are currently simply unaware.

We've probably only read the first page of the book of the cosmos, or at most the first chapter. There is a lot more yet to be learned from the rest of the volume.

Who knows what we might discover when we go out and have a look?

STARFLIGHT AND SPECIES MATURITY

In this chapter, we have discussed interstellar travel using mighty systems such as thermonuclear fusion, antimatter rockets, and laser-pushed light sails, all with power ratings in the tens of gigawatts to hundreds of terawatts. It should be obvious to most readers that such systems will be

(a) expensive, and
(b) very dangerous in the hands of minors.

As far as expense is concerned, this will take care of itself. Star flight will not occur until humanity can afford it. But as we have seen, if humanity does develop into a healthy Type II civilization, our resources and power base will continue to expand at a rate that will make even the huge costs associated with interstellar colonization affordable within just a few centuries.

The issue of danger is different. Star flight requires the deployment of vast amounts of energy in compact form. Any system that can dispense such energies is implicitly a weapon of mass destruction with potentials far exceeding the early twenty-first century's nuclear arsenals.

This brings up an interesting point. Intelligent species, including our own, evolve from aggressive, predatory, highly competitive forebears. It is, in fact, the selection pressures associated with the successful implementation of such a mode of life that call forth the evolution of that adaptation known as intelligence. Moreover, within the history of the species, it is the winners of millennia of tribal conflicts who survived to pass on their genes. It is a nasty but true fact that all of us alive today are descendants of folks who were good at killing people and breaking their stuff. We can be proud of it, we can be ashamed of it, but no matter how we feel about it, we are all the children of warriors.

Wars fought with bows, arrows, and spears are one thing; wars fought with antimatter bombs and planet-frying lasers are quite another. Primitive warfare can be very ugly, but it also carries the redeeming virtue of species selection for intelligence, social cooperation, and physical strength. Not so modern war. It is said that Christopher Columbus, observing a Spanish peasant disintegrate a squadron of proud Moorish cavalry with a hand-thrown bomb at the siege of Granada in 1492, commented that "such inventions make war meaningless." The author of humanity's Type I triumph saw far.

We are the children of warriors—but also of loving parents, incurable tinkerers, explorers, and reasoners. We bear the genes, instincts, and capabilities of all of these. From the warriors, we have inherited not only the instincts that threaten us but the courage to try the unknown. From the explorers, we have inherited the drive to take us to the stars; from the tinkerers, the spirit that will give us the tools to get there; and from the lovers and reasoners, we have received that which will allow us to use our expanding powers for good instead of evil.

There is no turning back. The spirit of the tinkerers and explorers cannot be suppressed without destroying our humanity. Safety at the cost of doing so would come at too high a price. So there will be Type II civilization with the capability to move asteroids, and there will be mass-produced plasma explosives, and giant lasers, and all the rest of the formidable gear needed to launch interstellar missions. And we will survive the test of their ownership. Because we also have Love and Reason, and when forced to do so, we can and will grow those capacities too.

People will do anything to survive, even become better.

The heavens will be open to those who deserve them. I think humanity does. We'll need to grow up a bit, but with ingenuity, determination, and species maturity, the stars can be ours.

FOCUS SECTION: NOAH'S ARK EGGS

Sending thousand-ton interstellar spaceships on half-century cruises at speeds of 10 percent the speed of light to nearby stars at $2 trillion each seems like a rather inefficient way to spread human civilization to the cosmos. The number of people each ship could take would be on the small size—perhaps a hundred or so, assuming that a multi-terawatt propulsion system could be built with a mass substantially less than one thousand tons—and to keep the trip time down to less than fifty years, the range could be at most five or six light-years, after which several centuries might be required to develop a civilization in the destination solar system capable of mounting expeditions to take the next step further. Proceeding in such a costly fashion, we might be able to reach the stars, but only with a net migration rate that might take us several millennia merely to settle our immediate hundred-light-year-radius interstellar neighborhood.

Visionary scientist Freeman Dyson has proposed an alternative: sending seed spaceships, which he calls Noah's Ark Eggs.[15] As Dyson explains it:

> The Noah's Ark Egg is a way of making space colonies highly cost-effective. They're very cheap, and also very powerful. They're using miniaturization to spread life in the universe, not just for exploring.
>
> The Noah's Ark Egg is an object looking like an ostrich egg, a few kilograms in weight. But instead of having a single bird inside, it has embryos—a whole planet's worth of species of microbes and animals and plants, each represented by one embryo.

It's programmed then to grow into a complete planet's worth of life. So it will cost only a few million dollars for the egg and the launch, but you could have about 1,000 human beings and all the life support, and all the different kinds of plants and animals for surviving. The cost per person is only a few thousand dollars, and it could enlarge the role of life in the universe at an amazingly fast speed. So you could imagine doing this in 100 years or so.

We'd need to know more about embryology before we could do it. We'd want to know how to design embryos, and design robot nannies to take care of them until they're grown up. But all that could be done.

This is a very intriguing idea. Let's consider how it might work.

As chapter 9 will discuss, we will soon have space telescopes capable of detecting oxygen in atmospheres of extrasolar planets, and a mature Type II civilization will certainly have much better instruments, capable of imaging such worlds and identifying spectral signatures of chlorophyll and other evidence of life. For reasons explained in chapter 9, I believe we will find many such worlds, creating a target-rich environment for such colonizing efforts.

That said, how do we get the arks to their targets? As I see it, the most promising approach would be to create a delivery system combining a laser-pushed light sail of the type being pioneered by Breakthrough Starshot for acceleration with a magnetic sail or other kind of plasma drag device for deceleration.

So let's run the numbers. Our interstellar spaceship has a mass of one kilogram, including a five-hundred-gram ark containing millions of fertilized eggs of numerous terrestrial life-forms, including humans. The other five hundred grams is a light sail with an average areal density of 0.01 g/m^2 (0.01 microns thick, equivalent), which would imply a diameter of 250 meters.

If we illuminate this sail with laser light with an intensity equal to a hundred times sunlight on Earth, the total amount of power required will be 6.75 gigawatts, the thrust will be 45.5 newtons, and the rate of acceleration will be about 0.46 g's. If we were to accelerate at this rate for six months, we will reach 23 percent the speed of light, while traveling a distance of about 8,200 AU (or 1.23 trillion kilometers).

Now, light spreads out as it goes, even when aimed by a large parabolic dish. The amount of spreading is given by the diffraction equation. If R is the range, λ is the wavelength of the light, S is the spot size at the target, and D is the diameter of the projecting dish, the relationship is given by:

$$R\lambda = (\pi/2)DS \qquad\qquad (7.1)$$

If we are using visible light with a wavelength of 0.5 microns and the spot size S is 250 meters, this reduces to:

$$R = (785{,}000{,}000)D \qquad\qquad (7.2)$$

So, for example, if D equals fifty kilometers, then R equals thirty-nine billion kilometers, or 260 AU.

But we have to keep light focused on the sail for 8,200 AU. If we did this with a single projector dish, it would need to be 1,600 kilometers in diameter, which is about half as big as the moon. That would be quite an engineering challenge. So rather than doing that, we use fusion-powered spacecraft to deploy a set of thirty-two laser relay stations, each with a fifty-kilometer dish, strung out over a path 8,200 AU, or 13 percent of a light-year, long. It will be recalled that fusion-powered ships could be expected to reach speeds on the order of 10 percent the speed of light, so they could travel out to the maximum relay station distance in a year or two. The ships themselves will use their fusion engines to provide the power to drive the lasers. Assuming these are about 50 percent efficient, each ship will need to generate about thirteen gigawatts of power in its turn.

Since the duration of the acceleration is six months, a total of 6.5 gigawatt-years of power will be needed to do the job, about what the city of Chicago consumes yearly. At current prices, that much power would cost about $3.4 billion, which would even be affordable today, let alone in the presumably much richer twenty-third century.

So now the ark is its way, at a speed on the order of a fifth that of light, able to reach Alpha Centauri in twenty years or Tau Ceti in about fifty. When it approaches its destination, it activates its plasma sail to create drag against the interstellar medium, thereby bringing its speed down to interplanetary-class velocities. This achieved, the craft could use its large light sail to navigate the destination solar system, getting it into orbit around the target planet. Then the little five-hundred-gram ark uses its two-hundred-gram, 0.2-square-meter heat shield (for a ballistic coefficient of 2.5 kg/m². For comparison, the SpaceX Dragon has a ballistic coefficient of 400 kg/m².). to safely enter the atmosphere and land its three-hundred-gram payload section on the ground.

Here we face a new challenge. The payload spacecraft has thousands of fertilized human eggs onboard, but human embryos need a womb to develop in and parents or parental surrogates to raise and teach them. We can't send either people or human-sized robots in a three-hundred-gram ark, and inorganic robots cannot be grown from tiny seeds. But plants and animals can.

There are, or were at one time, large, highly complex organisms on Earth, such as octopus, fish, and extinct amphibians, born from milligram-sized eggs. Most of these animals had no parental instruction but had all their required intelligence and necessary behaviors programmed into their genomes. So let us say that we made it our business to engineer a new species of such animals, perhaps resembling large salamanders like the newts in Karel Čapek's famous novel *The War with the Newts*, but programed with a set of instinctual behaviors allowing them not only to hatch, grow, survive, and multiply, but to build houses, farms, orchards, industries, incubators, nurseries, and schools suitable for gestating and raising the first generation of humans drawn from the thousands of fertilized eggs contained within the ark.[16] The ark is also equipped with a memory card containing all the principal technical knowledge and literature of human civilization so that as the human children grow, they can reacquire their intellectual heritage and then, aided by their newt assistants, proceed to build a new civilization on a new world.

Proceeding in such a way, human civilization could conceivably expand into interstellar space at a speed exceeding 20 percent the speed of light. Taking into account time needed to settle new solar systems along the way, by the year 3000, the human domain could extend more than one hundred light-years outward in all directions, encompassing a realm of more than twelve thousand solar systems.

There are many, many technical challenges to realizing the above scenario, so view it as wild speculation if you like. But it seems to me that if viewed from the standpoint of the probable capabilities of twenty-third-century science, there need be no showstoppers. The fusion-powered ten-gigawatt laser stations are mere extensions of known physical principles, operated on a scale that should be readily affordable by a mature Type II civilization. (We have many one-gigawatt fission power plants today.) It is already anticipated by many that the twenty-first century will see rapid advances in the fields of artificial intelligence and bioengineering, so developing the means to encode behaviors into animals—just as nature already does—to create self-growing and self-replicating robots should be well within our grasp by the end of the century. But it does raise some ethical questions.

In creating beings like the above-described newts, would we be reviving slavery? That's difficult to say. On the one hand, we clearly would be exploiting them. On the other, they wouldn't even exist without us. So in that sense, the relationship could fairly be described as symbiotic. We do need them if we are to colonize space using microspacecraft, because only animals that have all their necessary knowledge and behaviors programmed into them genetically can grow to maturity without the aid of parents. But even if they were made to physically resemble humans, with the capability of speech—which they (or robots that they in turn construct) would need to have to be able to raise sane children—would such preprogrammed beings actually be human? Humans don't know everything at birth. In fact, human babies are arguably the most ignorant of animals, totally outclassed in ability and sophistication by puppies and kittens, to say nothing of baby turtles or fish, which enter the world knowing everything they need to know—and nearly everything they will ever know. Humans can not only learn, but develop new mental abilities as they grow. That is why we need others to learn from. It is why we need to love and be loved. It is a big part of what makes us human.

So by that standard, the newts would not be human, and while treating them with cruelty would be wrong, giving them a life limited to their fixed abilities would not be. Moreover, creating allied species to help humankind has a long history. In fact, it has been essential to the success of humanity on Earth.

Twenty thousand years ago, our distant ancestors lived desperate lives. We had a hard time finding the game upon which our own lives critically depended, particularly in the harsh Ice Age winters, and were hunted ourselves by silent and powerful carnivorous cats who could easily take any human by surprise and tear him or her to pieces. But then, some clever and kind person—I would bet a little girl—had the wisdom to save a captured wolf pup and raise it as her own. Thus we created our friends, the dogs, who have been with us ever since. For our Stone Age ancestors, they were a godsend, providing game-tracking and early warning services essential to our survival. Freedom from want, freedom from fear: these two of the famous four freedoms, and with them our rise to success on Earth, were first made possible by our ability to forge a grand alliance with canine comrades.

If we wish to inherit the stars, we may need to pull the same trick again.

Figure 7.6. To the stars, people! Let's go! *Image courtesy of Kepler and Hope Zubrin.*

Chapter 8

TERRAFORMING: BRINGING NEW WORLDS TO LIFE

I live in Colorado, near the mountains. A few years ago, I had occasion to hike to the top of one of the smaller peaks, whose summit was just above the tree line. As I sat, observing the scenery while eating my lunch, an odd question crossed my mind: *How did all these trees get up here?* Conifers lined the slopes nearly to the summit of every peak in sight. *How did a mob of immobile trees ever climb those steep heights?*

As I munched away, pondering this, I noticed a group of chipmunks, scurrying about, *carrying pine cones*. The answer was thus made apparent. The chipmunks had transported the seeds uphill. Interesting. Every mountain in sight was covered with trees. By moving the seeds upslope, the chipmunks had enormously expanded their "natural habitat." In fact, if by "natural habitat" one means the habitat that would support a chipmunk population that exists prior to and independent of their seed-spreading activity, it's unclear whether any such place exists at all. The chipmunk habitat does not exist "naturally." It exists because the chipmunks (together with a host of other participating species) have *created* it. That's how life works.

A major challenge that humans will face as we become a Type II and then Type III civilization is that of transforming the environments found on other planets to more Earthlike conditions. This must be done because environments friendly to life are a product of the activity of life. Thus, as humans move out into space, it is unlikely that we will find environments that perfectly suit our needs. Instead, as life and humanity have done historically on Earth, we will have to improve the natural environments we find to create the worlds we want. Applied to other planets, this process of planetary engineering is termed "terraforming."[1]

Some people consider the idea of terraforming other planets heretical—humanity playing God. Others would see in such an accomplishment the most profound vindication of the divine nature of the human spirit—dominion over nature, exercised in its highest form to bring dead worlds to life. Personally, I prefer not to consider such issues in their theological form, but if I had to, my sympathies would definitely be with the latter group. Indeed, I would go further. I would say that failure to terraform constitutes failure to live up to our human nature, and a betrayal of our responsibility as members of the community of life itself.

These may seem like extreme statements, but they are based in history, about four billion years of history. The chronicle of life on Earth is one of terraforming—that's why our beautiful blue planet is as nice as it is. When the Earth was born, it had no oxygen in its atmosphere, only carbon dioxide and nitrogen, and the land was composed of barren rock. It was fortunate that the sun was only about 70 percent as bright then as it is now, because if the present-day sun had shined down on that Earth, the thick layer of CO_2 in the atmosphere would have provided enough of a greenhouse effect to turn the planet into a boiling, Venus-like hell. Fortunately, however, photosynthetic organisms evolved that transformed the CO_2 in Earth's atmosphere into oxygen, in the process completely changing the surface chemistry of the planet. As a result of this activity, not only was a runaway greenhouse effect on Earth avoided, but the evolution of aerobic organisms, which use oxygen-based respiration to provide themselves with energetic life styles, was enabled (though a primeval Environmental Protection Agency dedicated to preserving the status quo on the early Earth might have regarded this as a catastrophic act of environmental destruction). This new crowd of critters, known today as animals and plants, then proceeded to modify the Earth still more—colonizing the land, creating soil, and drastically modifying global climate. Life is selfish, so it's not surprising that all the modifications that life has made to the Earth have contributed to enhancing life's prospects, expanding the biosphere, and accelerating its rate of developing new capabilities to improve the Earth as a home for life still more.

Humans are the most recent practitioners of this art. Starting with our earliest civilizations, we used irrigation, crop seeding, weeding, domestication of animals, and protection of our herds to enhance the activity of those parts of the biosphere most efficient in supporting human life. In so doing, we have expanded the biospheric basis for human population, which has expanded our numbers and thereby our power to change nature in our interest in a continued cycle of exponential growth. As a result, we have literally remade the Earth into a place that can support billions of people, a substantial fraction of whom have been sufficiently liberated from the need to toil for daily survival that they can now look out into the night sky for new worlds to conquer.

It is fashionable today to bemoan this transformation as destruction of nature. Indeed, there is a tragic dimension to it. Yet it is nothing more than the continuation and acceleration of the process by which nature was created in the first place.

Life is the creator of nature.

Today, the living biosphere has the potential to expand its reach to encompass a whole new world, on Mars, and the Type II interplanetary civilization that develops as a result will have the capability of reaching much further. Humans, with their intelligence and technology, are the unique means that the biosphere has evolved to allow it to blossom across interplanetary and then interstellar space. Countless beings have lived and died to transform the Earth into a place that could give birth to a species with such capabilities. Now it's our turn to do our part.

It's a part that four billion years of evolution has prepared us to play. Humans are the stewards and carriers of terrestrial life, and as we spread out, first to Mars and then to the nearby stars, we must and shall bring life to many worlds, and many worlds to life.

It would be unnatural for us not to.

MARS-LIKE WORLDS

Mars is the first extraterrestrial planet that will be terraformed. As discussed in detail in my earlier book, *The Case for Mars*, the engi-

neering methods by which this can be done are relatively well understood. The first step will be to recreate the atmosphere of early Mars
by setting up factories to produce artificial greenhouse gases, such as
perfluoromethane (CF_4), for release into the atmosphere. If CF_4 were
produced and released on Mars at the same rate CFC gases are currently being produced on Earth (about a thousand tons per hour), the
average global temperature of the Red Planet would be increased by
10°C within a few decades. This temperature rise would cause vast
amounts of carbon dioxide to outgas from the regolith, which would
warm the planet further, since CO_2 is a greenhouse gas. The water
vapor content of the atmosphere would vastly increase as a result,
which would warm the planet still more. These effects could then be
further amplified by releasing methanogenic and ammonia-creating
bacteria into the now-livable environment, as methane and ammonia
are very strong greenhouse gases. The net result of such a program
could be the creation of a Mars with acceptable atmospheric pressure
and temperature, and liquid water on its surface within fifty years of
program start. Even though such an atmosphere would not be breathable by humans, this transformed (essentially rejuvenated) Mars
would offer many advantages to settlers: they could now grow crops
in the open; space suits would no longer be necessary for outside
work (just breathing gear); large supplies of water would be much
more accessible; aquatic life could flourish in lakes and ponds oxygenated by algae; and city-sized habitation domes could be constructed,
as there would be no pressure difference between their interior and
the outside world. These short-term advantages would be more than
sufficient to motivate Martian settlers to initiate the required terraforming operations to obtain them. In the longer term, plants spreading
across the surface of such a partially terraformed Mars would put
oxygen in its atmosphere. Using the most efficient plants available
today, it would take about a thousand years for enough oxygen to be
released to create an atmosphere that humans can breathe. However,
future biotechnology may allow the creation of more efficient plants,
or other technologies might become available which could accelerate
the oxygenation of Mars considerably.

It may be observed that if humans had encountered Mars not in its current condition but in its warm, wet youth, terraforming would have been much simpler. Essentially, we could have skipped the large-scale industrial engineering required to greenhouse the planet with CF_4, and gone directly to the stage of using self-replicating systems such as bacteria and plants to further warm and oxygenate the planet. Thus, a team of interstellar explorers who chance upon a "young Mars" could initiate massive terraforming operations with little more equipment than an appropriate array of bacterial cultures, seeds, and a bioengineering lab. (See plate 14.)

Mars-like and young-Mars-like worlds may be quite common in nearby interstellar space. If so, then the terraforming of Mars will thus give us not only one new world but the tools for creating many.

In January 2019, I received an email from a nurse named Patte, who asked how I thought it could be moral to terraform Mars when we have done so much damage to the Earth. Here is my reply.

Patte,

Mars is a dead or nearly dead world, with at most some microbes living deep underground.

I'm sure you would agree that it would be a horrible act of destruction to turn our own glorious living Earth into a dead world like Mars.

That being the case, then you must see what a wonderful act of creation it would be to transform a dead world like Mars into another glorious living world like Earth.

Furthermore, just as a doctor who chooses not to save a life when he could because he has problems of his own is acting immorally, so failing to bring Mars to life when we could would also be deeply immoral.

Humans are not the enemies of life. Humans are the vanguard of life.

Life to Mars, and Mars to life!

That's the wonderful calling that Nature has given us.

All the best,
Robert

WORLDS TOO HOT

Within our own solar system, the other planet that has attracted serious attention from would-be terraformers is Venus.

Venus was once thought to be Earth's sister planet, since it is about 95 percent the diameter of the Earth and has about 88 percent of Earth's gravity. The fact that it orbits the sun at 72 percent of Earth's distance implied that it would be warmer than Earth, but not necessarily fatally so, and visions of Venus as a world rich with steaming jungles beneath cloudy skies filled astronomy books through the late 1950s. However, in the early 1960s, NASA and Soviet probes reached Venus and discovered that the fair planet of the love goddess not only lacked jungles, but was a pure hell, sporting a mean surface temperature of 464°C—hot enough to melt lead. This was especially surprising, since Venus is masked by highly reflective clouds—so reflective, in fact, that the planet actually absorbs less solar radiation than the Earth! Based on the amount of sunlight it absorbs, Venus should be colder than Canada. Instead, it is as hot as a self-cleaning oven. The explanation for this paradox was soon found, however; Venus is hot because what heat it does absorb is kept captive due to the "greenhouse effect" caused by its thick CO_2 atmosphere. (This is how the "greenhouse effect" phenomenon currently of concern on Earth was first discovered.)

Well then, if the CO_2 atmosphere is baking Venus to death, why not just get rid of it? This was the genesis of Carl Sagan's seminal 1961 proposal to terraform Venus with aerial algae.[2] According to Sagan, Venus could be cooled by dispersing photosynthetic organisms in its atmosphere that could convert it from greenhousing CO_2 to transparent (and breathable) oxygen. This proposal was a landmark, in that it was the first serious discussion of terraforming within the world of science and engineering, as opposed to in science fiction. However, it would not have worked for a number of reasons.

In the first place, while algae have been found in rainwater, there are no plants that actually live in an aerial habitat. Perhaps they could be engineered, especially for a planet with as thick an atmosphere as Venus. Sagan's proposal faces a bigger problem, however. Though based on the scientific knowledge available at the time, his concept grossly overestimated the amount of water available on Venus. Photosynthesis involves combining water molecules with CO_2 molecules in accordance with:

$$6H_2O + 6CO_2 \Rightarrow C_6H_{12}O_6 + 6O_2 \qquad (8.1)$$

As can be observed in equation 8.1, to get rid of a molecule of CO_2 using photosynthesis, you need to use up a molecule of water. On Venus today, there is much more CO_2 than water, so if you could find organisms to perform reaction 8.1 on a mass scale, you would simply rid Venus of the small amount of water it retains while leaving the large bulk of the CO_2 atmosphere basically untouched. It would take the equivalent of a global ocean two hundred meters deep to provide enough water to react away the CO_2 in Venus's atmosphere via photosynthesis; in fact, Venus only has enough water to cover itself with a layer five centimeters deep. Sagan's idea could not have worked because there just isn't enough water on Venus to do the job.

Importing the water isn't an option. It would require moving ninety-two million iceteroids, each with a mass of a billion tons. So, in fact, the only way to cool Venus is to block the sun with a huge solar sail. If we were to make a sail twice the diameter of Venus out of aluminum 0.1 microns thick, 124 million tons of processed materials would be required, to which we would need to add several billion tons of ballast. Manufacturing this object out of asteroidal material would be a huge job, but probably not beyond the means of an advanced Type II civilization. Such a sail could be stationed at the position between Venus and the sun where their gravitational fields, sunlight force, and heliocentric centrifugal force all balance (near the Venus-sun L1 point, about a million kilometers from Venus). An alternative approach might be to put thick clouds of dust in orbit around Venus, thereby blocking sunlight. The advantage of this approach is that simple unprocessed material could be used. The disadvantage would be that a lot more mass would be required, and orbital operations could be seriously impaired for some time. In either case, we would aim to block more than 90 percent of the sun's rays, leading to the precipitation of Venus's CO_2 atmosphere as dry ice after a cooling-off period of about two hundred years. The dry ice could then be buried, and by moving the sail around in a small orbit about the L1 point, we could create an acceptable surface temperature and day/night cycle on Venus. But the planet would still be incredibly dry. (In contrast to Venus's 0.05 meters of water, Mars has 200 meters and Earth 2,000 meters. Only the moon, at 0.00003 meters, is dryer.) All this would tend to imply that terraforming Venus is a very big project offering modest payoff.

However, this was not always so. The young Venus had lots of water, comparable, in fact, to the water inventory of the Earth. According to the "moist greenhouse" theory proposed by planetary scientist James Kasting in 1988 and now generally accepted, the early Venus featured oceans of water at temperatures between 100° and 200°C. The pressure of the thick overlying atmosphere prevented these oceans from boiling away. The rapid cycling of water in this ultratropical environment would have caused most of Venus's CO_2 supply to rain out and react with minerals to form carbonate rocks. This moist greenhouse Venus could exist because the early sun was only about 70 percent as luminous as the sun is today.[3] But after a billion years or so, the sun's luminosity increased to 80 percent of its present value, and temperatures on Venus rose above 374°C, the critical temperature for water. Once this happened, liquid water could no longer exist on Venus, and all of its oceans turned to steam. With so much water vapor in the atmosphere, water loss from the planet due to ultraviolet dissociation of upper atmospheric water molecules occurred at a rapid rate. Moreover, once it stopped raining, geologic recycling was able to release the vast supplies of CO_2 stored in Venus's carbonate rocks, thereby creating the hellish runaway greenhouse environment that curses Venus today.

If human explorers had arrived on the scene when Venus was still in its young, moist greenhouse phase, terraforming could have been accomplished by building the sun shade described above. That would be a significant engineering project, but well worthwhile, since the result would be an Earth-sized planet complete with temperate oceans and a moderate-pressure, nitrogen-dominated atmosphere, fully ready for the rapid propagation of life.

As discussed in *The Case for Mars*, solar sails used as reflectors to increase solar flux will also be a useful auxiliary technique in melting the permafrost and activating the hydrospheres of cold Mars-like planets. A solar sail mirror fifteen thousand kilometers in radius (190 million tons of 0.1 micron aluminum) positioned with Titan at its focal point would increase Titan's solar flux to Mars-like levels, which, together with the strong greenhouse effects produced by Titan's methane atmospheric fraction, should be enough to raise the planet to Earthlike temperatures. A similar mirror would be sufficient to melt surface water ice into oceans and vaporize dry ice to create a CO_2 atmosphere on Callisto. Thus, the ability to manufacture large, thin solar sails in space, a central skill for engineering both Type II interplanetary transportation and emergent Type III interstellar spaceship propulsion, is also a key technology for terraforming Mars-like, Titan-like, Callisto-like, and young-Venus-like moist greenhouse worlds.

It should be noted that while Sagan's proposal for terraforming Venus with algae would not have worked, there is a class of planet for which it would: young Earths. The early Earth also had a thick CO_2 atmosphere, which was fortunate because the early sun was weak. But there could be many other young Earths out among the stars that are receiving solar input comparable to what the Earth gets today, but which are too hot to be habitable because of heavy CO_2 atmospheres. In the case of such worlds, appropriately selected or bioengineered photosynthetic organisms might be able to rapidly terraform the planet to fully Earthlike conditions without any additional macroengineering effort. This is important, because the Earth has only had its current high-oxygen/low-CO_2 atmosphere for the past six hundred million years, or the most recent 14 percent of the history of the planet. If humans had encountered the Earth during the other 86 percent of its history, we would have needed to adopt Sagan-like terraforming strategies in order to make it habitable.

So, Sagan didn't really have the wrong idea—he had the right idea but applied it to the wrong world. There are undoubtedly many right worlds out there waiting for its application. Based on our own history, young Earths, uninhabitable but ready for Sagan-style biological terraforming, are likely to outnumber already habitable Earths by a considerable margin.

ENGINEERING THE EARTH

The ability of humans to change planetary environments can be demonstrated by the effect we are already having on the Earth. Considering this necessarily brings us into a discussion of climate change, a scientific issue that unfortunately has been corrupted by political actors seeking to obscure, cherry-pick, or exaggerate various aspects of the case for partisan purposes. Nevertheless, despite the fact that any objective discussion of this matter is sure to evoke outrage by militants hailing from both wings of the spectrum, I will try to discuss this important matter in as level a manner as possible.

First of all, global warming is quite real. Indeed, it is demonstrable that the Earth has been warming for the past four hundred years. During Elizabethan times, the Thames River used to freeze over every winter, creating a place for "frost fairs" and other festivities. The last such fair was held in the mid-1600s, but as late as the mid-1800s, Charles Dickens could describe snowy winters in London, which no longer exist. During the American Civil War, Confederate soldiers stationed as far south as Georgia amused themselves by holding massive snowball fights pitting one regiment against another. By the early twentieth century, such climates were things of the past. Given the limitations of human industry over the period from 1600 to 1900, it's pretty clear that the warming that occurred during that span was natural rather than anthropogenic. That said, in the twentieth century, the rate picked up in a manner that is consistent with the predictions of the more conservative climate models for CO_2-driven global warming. On the basis of the available data, it would appear that coincident with a 33 percent rise of atmospheric CO_2 levels from three hundred to four hundred parts per million (ppm) over the past century (which is consistent with human fossil fuel use), global temperatures have risen an average of about 0.8°C. This result may be doubted by some because measuring an average global temperature to that degree of accuracy is a difficult task, with the results easily influenced by the choice of location for the thermometers. Nevertheless, it is clear that substantial warming has occurred because the average length of the growing season (i.e., the time between that last killing frost of the spring and the first one of the fall, an easy measurement to make) has expanded markedly over this period. For example, as shown by the EPA data presented in figure 8.1, the average length of the growing season in the United States has expanded by about twenty days since 1910, which is about as clear a demonstration of warming as you can get.[4]

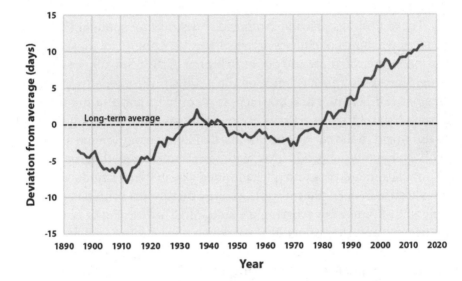

Figure 8.1. The length of the growing season in the United States has expanded markedly since 1895, a clear proof of climate change. *Image courtesy of the US Environmental Protection Agency.*

Unfortunately, since this is a beneficial change, those attempting to make the case for a climate emergency never mention it, leaving their argument hanging on unconvincing claims of statistically averaged precision temperature measurements.

Global warming should also lead to greater net rainfall *on average*, and this indeed has also been measured, although the degree of change varies locally, with some regions actually experiencing drought despite an overall increase in the total.[5]

In addition to driving warming, the easily measured and clearly anthropogenic 33 percent increase in atmospheric CO_2 is also having a powerful direct effect on the biosphere, with rates of plant growth worldwide accelerated by about 20 percent. This result, which is consistent with the well-established theory of photosynthesis, is supported by innumerable lab studies, field studies, and, most strikingly, satellite observations, as presented by the NASA data in plate 15 in the photo insert, which show a dramatic increase in the rate of plant growth over the past thirty-six years.[6]

(See plate 15.)

These findings provide grounds for dismissing some of the more strident warnings of climate activists. But let's not be too hasty. While the moderate warming and CO_2 enrichment of the atmosphere we have experienced *thus far* has been, on the whole, rather beneficial, *unconstrained* temperature and CO_2 increases could be another matter altogether. Much larger increases are possible, as simply raising the rest of the world to the *current* US standard of living would require quintupling global energy production, with even more needed in reality due to population increase and the continued rise of living standards in the advanced sector. Furthermore, the happy picture presented in plate 15 only shows what is occurring on land. Most of the Earth is covered by oceans, and they show little evidence of enhanced biological productivity due to increased CO_2. On the contrary, significant damage to coral reefs appears to be occurring due to CO_2-driven ocean acidification (although conventional pollution and overfishing may also be to blame in some cases).

So it would appear that massive anthropogenic CO_2 emissions are fertilizing the land while harming the oceans. This is happening because while CO_2 availability is a limiting factor for the growth of land plants, in most of the ocean the rate of growth of the phytoplankton that stand at the base of the food chain is controlled by the availability of trace elements, such as iron, phosphorus, and nitrates. This is why well over 90 percent of the biological productivity of the world's oceans comes from the less than 10 percent of their area that is fertilized with runoff from the coasts or continental shelf upwelling, with the vast open seas left a virtual desert. As a result, the ocean's CO_2 levels simply increase, with no useful and some potentially seriously harmful results.

What to do? The conventional answer from most of the global

warming activist community has been to propose increased taxes on fuel and electricity, thereby dissuading people of limited means from making much use of such amenities. This program seems to me to be both unethical and impractical, and regardless of anyone's opinion on the matter, it is quite clear that it is failing to impact the growth of global carbon emissions in any significant way.

The basis for a much more promising approach was demonstrated by the British Columbia–based Haida First Nations tribe, who in 2012 launched an effort to restore the salmon fishery that has provided much of their livelihood for centuries. Acting collectively, the Haida voted to form the Haida Salmon Restoration Corporation, financed it with $2.5 million of their own savings, and used it to support the efforts of American scientist-entrepreneur Russ George to demonstrate the feasibility of open-sea mariculture through the distribution of 120 tons of iron sulfate into the northeast Pacific to stimulate a phytoplankton bloom, which in turn would provide ample food for baby salmon.

By 2014, this controversial experiment proved to be a stunning, over-the-top success. In that year, the number of salmon caught in the northeast Pacific more than quadrupled, going from 50 million to 219 million. In the Fraser River, which only once before in history had a salmon run greater than 25 million (about 45 million in 2010), the number of salmon increased to 72 million.

"Up and down the West Coast fisheries scientists and fishers are reporting they are baffled at the miraculous return of salmon seen last fall and expected this year," commented George. "It is of course all because when we take care of our ocean pasture. Replenish the vital mineral micronutrients that we have denied them through our high and rising CO_2 just one old guy (me) with a dozen Indians can bring the ocean back to health and abundance."[7]

In addition to producing salmon, this extraordinary experiment yielded a huge amount of data. Within a few months after the ocean-fertilizing operation, NASA satellite images taken from orbit showed a powerful growth of phytoplankton in the waters that received the Haida's iron.[8] It is now clear that as hoped, these did indeed serve as a food source for zooplankton, which in turn provided nourishment for multitudes of young salmon, thereby restoring the depleted fishery and providing abundant food for larger fish and sea mammals as well. In addition, since those diatoms that were not eaten went to the bottom, a large amount of carbon dioxide was sequestered in their calcium carbonate shells.

Unfortunately, the experiment, which should have received universal acclaim, was denounced by many leading environmental activists. For example, Silvia Ribeiro, of the international environmental watchdog ETC group, objected to it on the basis that it might undermine the case for carbon rationing. "It is now more urgent than ever that governments unequivocally ban such open-air geoengineering experiments. They are a dangerous distraction providing governments and industry with an excuse to avoid reducing fossil fuel emissions." Writing in the *New York Times*, Naomi Klein, the author of a book on "how the climate crisis can spur economic and political transformation," said that "at first, . . . it felt like a miracle."[9] But then she was struck by a disturbing thought:

> If Mr. George's account of the mission is to be believed, his actions created an algae bloom in an area half of the size of Massachusetts that attracted a huge array of aquatic life, including whales that could be "counted by the score." . . . I began to wonder: could it be that the orcas I saw were on the way to the all-you-can-eat seafood buffet that had descended on Mr. George's bloom? The possibility . . . provides a glimpse into the disturbing repercussions of geoengineering: once we start deliberately interfering with the earth's climate systems— whether by dimming the sun or fertilizing the seas—all natural events can begin to take on an unnatural tinge. . . . A presence that felt like a miraculous gift suddenly feels sinister, as if all of nature were being manipulated behind the scenes.

But the salmon are back.

Not only that, but contrary to those who have denounced the experiment as reckless, its probable success was predicted in advance by leading fisheries scientists. "While I agree that the procedure was scientifically hasty and controversial, the purpose of enhancing salmon returns by increasing plankton production has considerable justification," Timothy Parsons, professor emeritus of fisheries science at the University of British Columbia, told the *Vancouver Sun* in 2012. According to Parsons, the waters of the Gulf of Alaska are so nutrient poor they are a "virtual desert dominated by jellyfish." But iron-rich volcanic dust stimulates growth of diatoms, a form of algae that he describes as "the clover of the sea." As a result, volcanic eruptions over the Gulf of Alaska in 1958 and 2008 "both resulted in enormous sockeye salmon returns."[10]

The George/Haida experiment is of historic significance. Starting as a few bands of hunter-gatherers, humanity expanded the food resources afforded by the land a thousandfold through the development of agriculture. In recent decades, the bounty from the sea has also been increased through rapid expansion of aquaculture, which now supplies about half our fish. Without these advances, our modern global civilization of seven billion people would not be possible.

But aquaculture makes use only of enclosed waters, and commercial fisheries remain limited to the coasts, upwelling areas, and other small portions of the ocean that have sufficient nutrients to be naturally productive. The vast majority of the ocean, and thus the Earth, remains a desert. The development of open-sea mariculture could change this radically, creating vast new food resources for both humanity and wildlife. Furthermore, just as increased atmospheric carbon dioxide levels have accelerated the rate of plant growth on land, so increased levels of carbon dioxide in the ocean could lead to a massive expansion of flourishing sea life—more than fully restoring the world's depleted wild fisheries—provided humans make the missing critical trace elements needed for life available across the vast expanse of the oceans.

The point deserves emphasis. The advent of higher carbon dioxide levels in the atmosphere has been a boon for the terrestrial biosphere, accelerating the rate of growth of both wild and domestic plants and thus expanding the food base supporting humans and land animals of every type. Yet in the ocean, increased levels of carbon dioxide not exploited by biology could lead to acidification. By making the currently barren oceans fertile, however, mariculture would transform this apparent problem into an extraordinary opportunity.

Such an effort would more than suffice to limit global warming. Indeed, were 3 percent of the Earth's open-ocean deserts enlivened by mariculture, the entirety of humanity's current CO_2 emissions would be turned into phytoplankton, with available worldwide fish stocks greatly increased as a result.[11]

The situation is ironic. In some places in the ocean, excessive nutrients delivered by runoff of agricultural fertilizers cause local algae blooms that are so massive as to destroy all other aquatic life. Yet when delivered in the right amounts, such "pollutants" become the key to creating a vibrant marine ecology.

You can irrigate a farm to make it productive, or you can flood it and destroy the crops. You can fertilize land with horse manure, or you can . . . well, you get the idea.

"Pollution" is simply the accumulation of a substance which is not being put to good use. Carbon dioxide emissions are neither good nor bad in themselves. They are good for parts of the biosphere that are ready to make use of them and bad for those that are not.

Say what you will, humans are not going to stop using fossil fuels any time soon. So we need to prepare the Earth to take full advantage of the resulting emissions.

We live on fertile islands in a saltwater desert planet. If we apply some creativity, we can make those deserts bloom.

ROBOTICS, BIOENGINEERING, NANOTECHNOLOGY, AND PICOTECHNOLOGY

Terraforming will require a lot of work, and the people who attempt it will want helpers. There is thus no doubt that in terraforming, as in all other extraterrestrial engineering projects, robotics will play an important role. No commodity will be in shorter supply in an early Martian colony than human labor time, and as the frontier moves outward among the planets and then to the stars, the labor shortage will grow ever more pressing. The space frontier will thus serve as a pressure cooker for the development of robotics and other forms of labor-saving technologies.

But robots that must be manufactured still demand human labor. This will make them expensive, as space labor will be dear and transportation from Earth will be costly. Expensive robots are acceptable for assisting in certain tasks, such as exploration, where large numbers are not required. But terraforming will need multitudes. The only solution would be robots that make themselves.

Back in the 1940s, the mathematician John Von Neumann proved that self-replicating automatons are possible. That is, he proved that there is no mathematical contradiction that precludes the existence of such systems. But creating them is another issue altogether.

No one today has a clue as to how to do it, but it would not be too big a leap of faith to believe that a machine could be built and programmed that, if let loose in a room filled with gears, wires, wheels, batteries, computer chips, and all its other component parts, could assemble a copy of itself. But who would make the parts? Consider what is necessary to make even a simple part, such as a stainless steel screw.

To make the steel for the screw, iron, coal, and alloying elements from all over the world need to be transported to a steel mill. They need to be transported by rail, ship, truck, or plane, and all of these contrivances must be made in factories or shipyards of great complexity, each of which involves thousands of components shipped in from all over the world, by various devices, made in various facilities, and so on. So just supplying the steel for the screw involves the work of thousands of factories and millions of workers. If we then consider who made the food, clothing, and housing used by all those workers, who taught them, and who wrote the books that educated them, we find that a large fraction of the present and past human race was involved. And that's just the steel for the screw. If we now consider the processes needed to put the thread on the screw . . . but I think you get my point. Self-replicating machines cannot exist unless the parts they require are ready-made. This will never be the case for machines built out of factory-produced gadgets.

The only self-replicating complex systems known to exist are living things. Organisms can reproduce themselves because they are made of cells that can reproduce themselves using naturally available molecules as parts for their component structures. Because they can reproduce themselves, bacteria, protozoa, plants, and animals have extraordinary power as terraforming agents; a few of the right kinds, released under the right conditions, can multiply exponentially and radically transform an environment. Of course, for the transformation to be beneficial, some aspect of the organism's self-directed activity must contribute to the terraforming program. As we have seen, in such cases as methanogenic bacteria producing greenhouse gases, or photosynthetic plants eliminating them (while producing useful oxygen), the metabolisms of many forms of life make them natural servants of the terraforming process. This is to be expected, because as discussed earlier in this chapter, life would not exist if it did not terraform.

That said, current bacteria, plants, and animals are not specifically adapted to terraforming virgin planets; their adaptations are focused on terraforming and living on the current Earth. Their ancestors pioneered the early Earth, and they retain some of the necessary skills, but they are by no means the ideal candidates for pioneering new worlds.

However, since the domestication of the dog, twenty thousand years ago, humans have practiced modification of other species to meet our needs, primarily through the practice of selective breeding. In recent years, a series of advances—first the development of genetics, then the discovery of DNA, and now the actual reading of the genetic code and mastery of recombinant DNA techniques—has enormously expanded our abilities in this area. As a result, it will soon be within our capabilities to design ideal pioneering microorganisms and ultraefficient plants well suited to transform a wide variety of extraterrestrial environments.

But microorganisms and plants have their limits. They are all based on water/carbon chemistry, which cannot function beyond the temperature boundaries defined by the freezing and boiling points of water. If temperatures are sustained below 0°C, life survives but goes dormant; above the boiling point (100°C at Earth's sea level, 374°C maximum), organisms are destroyed. Many extraterrestrial environments of interest exist beyond these narrow limits.

The question thus arises if it might be possible to develop self-replicating organisms with a fundamental chemistry other than the water/carbon type universal to life as we know it. If in venturing out into interstellar space we should discover novel kinds of life, based on silicon or boron, for example, but with their own equivalent of a genetic code that future human bioengineers can master, this problem would be partially solved, as the new chemistry would undoubtedly define a new set of temperature limits. But it is unclear whether such organisms will ever be found, or what the extent of their utility might be. From the point of view of the planetary engineer, a more intriguing question is whether we can devise from scratch self-reproducing organisms that are not water/carbon.

This is the idea behind "nanotechnology"—the construction of self-replicating microscopic programmable automatons out of artificial structures built to design specifications on the molecular level. Why try to build microscopic self-replicating robots when we don't even know how to build human-scale reproducing automatons? The reason, once again, is that parts for large robots need to be manufactured in advance, while the molecules used as parts for nanorobots (a nanometer is a billionth of a meter) either come ready-made or can be readily assembled from atoms that do. So while building a nanorobot would unquestionably be more difficult that constructing a normal-sized one, nanorobots are the only kind that hold the promise of being potentially self-replicating.

The vision of nanotechnology is described at length by the field's champion, K. Eric Drexler, in his book *Engines of Creation*.[12] The basic idea is that once we learn how to manipulate individual atoms and molecules, machines can be constructed using small clumps of atoms as gears, rods, wheels, and other parts. Because each of these parts would be made from a precisely assembled group of atoms—just as carbon is arranged in a diamond lattice—they potentially could be very strong. Thus nanomachines filled with gears, levers, clockwork, motors, and all kinds of mechanisms could, in principle, be built. Energy to drive the units could be obtained from nanophotovoltaic units and stored in nanosprings or nanobatteries. To go from there to nanorobots, we need nanocomputers. Drexler proposes that these could be built out of mechanical nanomachines, along the same principle as the first mechanical computers proposed by Charles Babbage and Ada Lovelace to be built out of brass gears and wheels in the nineteenth century. Such machines could be programmed with punched tape or cards, and presumably nanoscopic analogs for these mechanical software devices could be found as well.

Babbage's ingenious mechanical computers don't even remotely compare in capability to modern electronic ones, but the parts used by Drexler's nano–Babbage machines would be so small that enormous amounts of computing power could be contained in a microscopic speck. So, once we accomplish the admittedly difficult job of building the first nanorobot, with all its necessary nanomechanisms for loco- motion and manipulation, and equip it with a superpowerful version of a Lovelace-programmable Babbage machine built on the nanoscale, we could set this first "assembler" loose and it would multiply itself through exponential reproduction. The vast horde of assemblers would then turn their attention to accomplishing some task they had been programmed to execute, such as inspect a human body for cancer cells and make appropriate adjustments, manufacture huge solar sails from asteroids, or terraform a planet.

Figure 8.2. Hardware and software. Charles Babbage invented the mechanical computer. The mathematician Countess Ada Lovelace, daughter of the poet Lord Byron, realized that Babbage's computers could be programmed to act like mechanical brains. Her insight could enable self-replicating nanorobots, endowing humanity with nearly unlimited power to terraform worlds.

To build macroscopic structures, billions of nanorobots would have to group themselves together to form large robots, perhaps on the human scale or even much bigger. This could lead to the manifestation of systems that would have all of the capabilities of the evil "liquid metal" robot depicted in the movie *Terminator 2*, able to change its shape and disperse and reassemble itself as required. But it actually would be much more powerful, since when it did choose to disperse, each of the billions of its subcomponents could be used as a seed to reassemble an entire unit from dirt. Even Arnold Schwarzenegger would have had a hard time saving the world from one of those!

It certainly sounds like fantasy, but is it? In defense of the nanotechnology thesis, one can advance the statement that it does not defy any known laws of physics, and therefore, given sufficient technological advance, it should become possible. Against it, one can easily point out the enormous technological difficulties that must be mastered before nanotechnology becomes a reality. Furthermore, while nanotechnology may not violate any laws of physics, *controllable* self-replicating robots may well violate the laws of *biology*. Consider: small replicating nanomachines will unquestionably undergo random alterations, or mutations, if you will. Those mutations that produce strains that reproduce more rapidly will swiftly outnumber to insignificance those that don't. Clearly, if the goal is to rapidly reproduce, it would be to a nanomachine's advantage not to have to bother with doing work for the benefit of human masters. Instead, evolutionary pressures will dictate that nanorobots attend only to their own needs. Those nanorobots that continue to slave away in obedience to their human-directed programs will not be able to compete with the wild varieties and will rapidly go extinct. As the saying goes, "Live free or die."

There is another reason to hold inorganic nanorobots suspect—we don't observe them. If diamond-geared self-replicating assemblers could be built, they would be ideally suited for dispersal across interstellar space using microscopic solar sails for propulsion. If, in the vast sweep of past time, a single species anywhere in the Milky Way developed such micro-automatons, it long since would have been able to use them to colonize the entire galaxy. All life on Earth would be based on nanorobots. But since this is not observed, we are driven toward concluding that either (a) there is no other intelligent life in the galaxy, or (b) nonorganic nanotechnology of the self-replicating micro-Babbage robot type described by Drexler is impossible. Since we know that the

evolution of intelligent life is possible, but we do not know that nanotechnology is, I must consider (b) the more likely alternative.

It may be observed that bacteria, the organic nanocritters of
nature, are also capable of surviving spaceflight. We would therefore
expect that if bacteria had evolved (or been developed) elsewhere in
the galaxy, that they would be the basis for life on our planet. Interestingly, they are. Not only are bacteria the earliest known inhabitants
of the Earth, but the higher eukaryotic cells that compose all animals
and plants are clearly evolved from symbiotic colonies of bacteria. The
possible broader significance of this will be discussed further in a following chapter. For our purposes here, however, it suffices to say that
the omnipresence of organic self-replicating nano-spacefarers (bacteria) and the absence of nonorganic nano-assemblers are evidence
for doubting the feasibility of Drexler-style nanotechnology.

But maybe nanotechnology isn't impossible—maybe it's just incredibly difficult. Maybe the reason nobody else has invented it is because
they weren't smart enough, or didn't try long and hard enough, or were
scared of the consequences of it getting out of control. Maybe everyone
else just decided that using bacteria was easier and sufficed for their
purposes. Maybe there really is a way to initiate and control nanotechnology, and it's just waiting for someone to invent. In every field of
endeavor, someone has to be first. Maybe that someone could be us.

That's a lot of maybes. But it's worth some speculation, because if
the promise ever does pan out, programmable self-replicating nanomachines will offer our descendants powers of creation limited only by the
rate at which solar flux provides the energy needed to drive work in a
given region. If we continue the vector toward ever-growing technological sophistication that necessarily will accompany our transformation
into first a Type II and then Type III civilization, the intricate wizardry
required to develop nanotechnology might someday fall within our
grasp. And who knows? Perhaps, in the still more distant future, even
greater capabilities could become possible—building machines not
out of atoms or molecules but from subatomic particles such as atomic
nuclei. Operating on a scale thousands of times smaller and faster than
even nanomachines, such *picotechnology* might draw its energy not from
chemical reactions but from far faster and more powerful nuclear reac-

tions. The capabilities that such programmable picomachines would make available could only be described today as sheer magic.

In the meantime, however, my bet is on bioengineering. Life offers us a tried-and-true type of self-replicating micromachine, and the programming manual is already in our hands. With our brains and their muscle, human-improved microorganisms will do some very heavy lifting in the hard work required to bring dead worlds to life.

LIGHTING STARS

It is better to light a candle than to curse the darkness.

Stars are the sources of life. Enormous engines of nuclear fusion, they pour light out into the cosmos, warming the dead cold of space and providing the antientropic power needed for the self-organization of matter. Starry nights have a mystic beauty, but when considered from a scientific standpoint, they are even more beautiful than they look. For the million specks of light that adorn the black velvet of a dark night sky are, in fact, nothing less than a million fountains of life.

Without question, there are numerous worlds too far from any star to support life. In our own solar system, we find world-sized moons of Jupiter and Saturn that can only be terraformed with the aid of giant reflectors, and worlds beyond, such as Neptune's giant moon Triton, for which the huge efforts required for such an expedient make it difficult even to contemplate.

Our sun is actually among the brightest 10 percent of stars. Most of the stars we know are much dimmer type K or M red dwarfs, and there are likely legions dimmer still: brown dwarfs, too small to ignite fusion, whose dead planets therefore orbit endlessly in frozen darkness.

What if we could light their fires?

If the object in question is an actual luminous star, such as a type M red dwarf, or even a brown dwarf, we could amplify its power by using solar sails to reflect back a small portion of the star's output. The rate at which thermonuclear fusion proceeds in a star goes as a strong power of its temperature. For proton-proton fusion in a star the size and temperature of our sun, reaction rates scale as the fourth power of temperature, while for cooler stars, the temperature dependence is stronger. If the CNO cycle is being used by the star to catalyze fusion, then the reaction rate will increase in proportion to T^{20} (!!!). So even increasing the temperature of a star by a small amount through reheating with reflected light can cause a large increase in power generated. This increased output will cause the star's temperature to rise further, which will amplify output yet again.

Type M stars outnumber all other types three to one. Brown dwarfs may well outnumber all luminous stars put together by several orders of magnitude. If we can set these bodies alight, we can vastly expand the domain of life in the universe.

In his book *Star Maker*, written in the 1930s, the British philosopher Olaf Stapledon compared the star maker to God.[13]

Gods we'll never be. But starmaking *is* a very noble profession.

If we learn to light stars, we will become capable of bringing not only planets but whole solar systems to life. That's not too shabby for the children of tree rats.

In the early universe, nearly all matter was hydrogen or helium. The heavier elements, including the carbon, oxygen, and nitrogen vital to organic chemistry and life, were all made in stars and spread through the cosmos by stars in nova and supernova explosions. We are stardust, warmed to life by a mother star and now ready to leave the nest to seek our fortune and make our mark among her siblings.

Ex astra, ad astra.

Stars have made life. Life should therefore make stars.

FOCUS SECTION: THE KEPLER MISSION

Because not only the sun but almost every large object in our own solar system is encircled by smaller orbiting bodies, it has always been reasonable to suppose that the stars have planetary systems too. Indeed, as far back as the Renaissance, Italian philosopher Giordano Bruno advanced the claim that the stars were indeed suns, surrounded by worlds that, like our own, are inhabited by intelligent beings. These people when they look up into the sky see us, and therefore, "we are in heaven." In other words, the laws of the heavens are the same as those of the Earth. This being true, the human mind should be able to comprehend the nature of the universe.

For this daring hypothesis, which stands as the fundamental basis of science itself, Bruno was burned at the stake in 1600. Notwithstanding his fate, many others more favorably situated tried to prove it over the centuries that followed, using ever-better telescopes to search for extrasolar planets. However, it was not until the 1990s that a single one was detected.

The reason for this rather extended delay is that any planet orbiting another star would not only be billions of times dimmer than a planet in our own solar system—some of which, like Uranus, Neptune, and Pluto, were not themselves detected until fairly recent times—but outshone billions of times over by their own stars. This combination made them practically invisible, even to very good twentieth-century instruments like the two-hundred-inch-diameter Mount Palomar telescope.

With direct imaging nearly impossible, more sophisticated techniques were advanced. The first to succeed involved looking for the wobble that might be induced in a star's motion by a heavy planet orbiting close by, pulling the star this way and that as it circled about. If the planet's orbit lay close to the plane pointing toward the Earth, the star would be pulled toward us when the planet came in front, temporarily Doppler shifting its light toward the blue end of the spectrum. Similarly, when it went around the back, it would pull the star away, shifting its light toward the red. By observing these periodic shifts in spectra, the existence of such planets could be inferred. Using this technique, Pegasus 51b, the first extrasolar planet orbiting a normal star, was discovered in 1995, with several more found each year afterward. (An alternative gravity-induced-wobble technique enabled discovery of a planet orbiting a neutron star in 1992.)

However, this Doppler-wobble detection method only works for really massive planets orbiting very close (as in a few percent of the Earth's distance from our sun) to their stars and so is limited to finding uninhabitable "hot Jupiters." Our own solar system would be undetectable to alien astronomers if they were restricted to this technique.

There had to be a better way, and NASA Ames astronomer William Borucki thought he had it. Instead of looking for reflected light emitted by an orbiting planet, he reasoned, why not look for the much larger loss of light that would occur when a planet passed in front of its star, partially eclipsing it as seen by us? The Earth is a hundredth the diameter of the sun, and so it would eclipse a ten-thousandth of the sun's light as seen by an observer in another solar system. The same would hold looking the other way. By the 1980s, photometers with such sensitivity were available. If we could launch one into space, attached to a good telescope with appropriate gear for recording massive amounts of data to compare light levels observed from thousands of stars over time, we would be able to detect all kinds of planets, including other habitable Earths!

So Borucki began to campaign for such a mission, holding the first workshop to discuss the idea in 1984, to pull together a team. Progress was slow in convincing the NASA establishment, which had other priorities, but things began to open up in the 1990s when NASA administrator Dan Goldin initiated the Discovery program, which invited independent teams with creative mission concepts to seek funding in repeated open competitions, with proposals judged comparatively on the basis of cost, risk, and perceived scientific value. Borucki's team entered the competition and lost. They tried again the second time and lost again, and then again. But the fourth time around, they won. In 2001, the Kepler mission (so named after the discoverer of the laws of planetary motion, at the insistence of team members Carl Sagan and Jill Tarter) became the tenth mission to be funded by the Discovery program.

After much hard work, budget struggles, delays, and other heartache, in March 2009 Kepler was finally launched into interplanetary space, following behind the Earth about sixty degrees in its orbital track around the sun. Aiming its 0.95-meter Schmidt telescope toward a region in the sky between the stars Deneb in the constellation Cygnus and Vega in Lyra, it began its scientific reconnaissance in June 2009. Deneb and Vega are two of the three corners of the "Summer Triangle" well-known to stargazers (the third is Altair, in the constellation Aquila). They can be easily seen by observers in the northern hemisphere on a clear summer night. They lie in the Milky Way, but not its densest-packed part, which would be toward the center of our galaxy.

Figure 8.3. Kepler's field of view. *Image courtesy of NASA.*

Rather, they lie in the plane of our galaxy but in the direction of the sun's orbit around its center, so that when we look toward them, we are seeing stars that co-orbit the galaxy with us at our same distance from the galaxy's core—which is to say, about halfway out. The field of view of Kepler encompasses about ten degrees by ten degrees, allowing it to see about 0.25 percent of the sky at any one time. Within this field of view, there were about 500,000 stars visible to Kepler, which the team cut down to 150,000 candidates for observation, as many of the others had natural variations in brightness that would make Kepler's photometric technique of planet detection ineffective. The telescope then began to collect data and process it onboard to limit the load for transmission, and the astronomers began to watch and wait for the results.

A planet will only eclipse its star if it orbits it in the plane that points toward to Earth. This fact reduces its chance of detection by Kepler by a factor of two hundred. Furthermore, since Kepler requires three eclipses from a given planet to establish that a single body exists that performed two orbits of the same period while Kepler was watching, only planets with orbital periods shorter than half the total observation time can be detected. So, for example, after the first two months of operation, Kepler could only detect planets with orbital periods ("years," if you will) of less than one month. To go around so swiftly, a planet would have to orbit very close to its star—in our own solar system, it would circle about halfway between Mercury and the sun. So such superhot worlds are easiest to detect, and, to the joy of the team, Kepler found dozens of them virtually immediately. But the planets of greatest interest to us are those that could possibly harbor life, which orbit further out and, moving more slowly around a longer path, take longer periods of time to make their circuit, and thus to be detected. The longer Kepler could keep its eyes focused on its single field of view, the more interesting its results would become. But Kepler was a low-cost spacecraft built and operated within the limited budget of a Discovery mission: about $500 million, around a fifth as much as flagships like Galileo or Cassini. To find habitable worlds with Earthlike years, it would need to keep working well for at least three years. How long could it last?

It lasted four years. And not only did it detect many Earthlike worlds orbiting in their star's habitable zone, it has found innumerable worlds of every possible description, orbiting all over the place. It found rocky worlds and gaseous worlds, and water worlds, and ice worlds, and lava worlds, and iron worlds, and diamond worlds—solar systems galore with

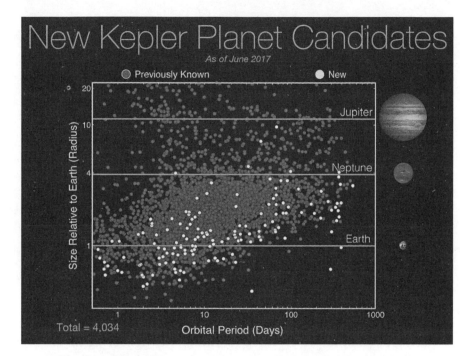

Figure 8.4. Kepler has discovered more than four thousand planets. The nature of the instrument causes bias toward finding larger planets with short orbital periods. Much larger numbers of smaller, long-period planets may be inferred. *Image courtesy of NASA.*

components and arrangements unanticipated even by science fiction writers.[14] In all, it detected some 4,000 candidate planets, with more than 2,500 now confirmed. These statistics are incredible, because given the limits of its detection technique, Kepler could only have been expected to find around 800 planetary systems among its 150,000 targets. The implication is that virtually all stars are surrounded by solar systems with multiple planets. That means that there are hundreds of billions of planets to be found in our own galaxy alone. Furthermore, Kepler threw in the trash can all previous theories of planetary formation based on explaining why our own solar system, with its four small rocky planets orbiting close to the sun and surrounded by four large gas giants orbiting further out, necessarily had to be constructed that way. In the words of one scientist: "It is now apparent that any planet which is physically possible exists."

We are living in a universe of all possible worlds.

Two of the four reaction wheels necessary for accurately pointing Kepler gave out in 2013, putting an end to its systematic long-duration survey of the Deneb-Vega sector. But the resourceful team came up with a clever idea of using natural forces to keep Kepler pointing away from the sun, allowing it to sweep through the zodiac with a series of sixty-day observing campaigns on each sector. This has allowed it to keep operating, discovering hundreds of additional short-period planets observable by looking outward through our own solar system's ecliptic plane.

The resounding success of Kepler inspired photometric planet detection campaigns to be launched by many large ground-based telescopes, resulting in the discovery of an additional 1,500 planets all over the sky. More powerful follow-up will soon be available from the recently launched TESS, which promises to multiply Kepler's discoveries a thousandfold.

But the summary result is already clear. As noted astronomers Sara Seager and Andrew Howard summed it up delivering the prestigious Pickering Lecture to the American Institute of Aeronautics and Astronautics in September 2018:

"On the basis of the available data, we estimate that 20 percent of stars have Earthlike planets in their habitable zone."

Wow.

Figure 8.5. Kepler team member Sara Seager delivers the astonishing mission verdict: one in five stars harbors an Earth-sized planet in its habitable zone.

Part II

WHY WE MUST

Chapter 9

FOR THE KNOWLEDGE

Many people believe our present scientific knowledge encompasses all that is important in the universe—that we now know essentially all there is to know. This conceit is wildly wrong, in fact nearly as wrong as the same conceit held by most people in previous ages. In fact, we know virtually nothing about the origin of the universe, its laws, and its principal components, including space, time, matter, life, and intelligence.

We really don't know that much about physics today. We understand *how* a variety of phenomena work, but we don't know *why*. We don't know why matter exists or has mass, or why mass has inertia or bends space-time to exert gravity. We don't know why all mass is positive (as opposed to negative or imaginary)—or if indeed it truly is. We understand how charges interact, but we don't know why charge exists, or why the charge of the fundamental particles are what they are, or why like charges repel but unlike charges attract, or why the magnitude of these forces have the particular values they do. We don't know why mass-energy or charge is conserved, or if it actually is under all circumstances. We don't know why fundamental particles with a given self-repelling charge don't blow themselves to pieces. We know that the ratio of electric and magnetic forces determines the speed of light, but since we don't know why these forces have the ratio they do, we really don't know why light travels at the speed it does. We don't really know what space, or time, or space-time, is, or where it comes from, where it might go, or why it is continuous. We don't know why there are four, and only four, fundamental forces of nature, or if indeed there are only four. We don't know what caused the big bang, or time, or causality for that matter. We don't know why time runs forward but not backward or sideways. We don't know if our universe is unique, or if millions or trillions or even infinite numbers also exist. By definition, if something is in another universe, it is not in ours, and we cannot interact with it. But was this always so, or must it always be so?

We don't know why the laws of the universe follow the geometric relations they do, or why they should have any relation to geometry at all. We don't know why the fundamental constants governing the magnitude of forces in our universe are constant, or if they truly are.

We are similarly ignorant of biology. We know about the life we see here on Earth, but we don't know how it started, why it chose the path it did, or what the true limits of its possibilities are. We suspect we are not alone, but we don't have a clue as to who else is around, what they are like, what they are up to, and what they know. These are only a few examples of our ignorance. The amount of things we have yet to learn is immense.

If answers to any of the above questions were found, the value could be extraordinary, enabling us to do things that are considered physically or medically impossible today. But to find these answers, we are going to leave our cradle planet and have a serious look around.

ASTRONOMY FROM SPACE

There is no better place to do astronomy than space. Our existing space observatories, such as the Hubble Space Telescope, have already revolutionized our knowledge of the universe. Indeed, Hubble discovered dark energy, which comprises 70 percent of the substance of the cosmos, and which is responsible for a fifth force of nature (beyond the previously known gravitational, electromagnetic, strong nuclear and weak nuclear forces) that is causing the expansion of the universe to accelerate. So while scientists before Hubble's 1990 launch thought they pretty much knew the nature and laws of the universe, without the space telescope, we today would be completely unaware of most of what exists, and what all that exists is mostly doing. In addition, Hubble made numerous other discoveries bearing on the origin of planets, stars, galaxies, and quasars, among other things, not to mention providing, through its spectacular images, some of the greatest and most uplifting art of modern times.

The Kepler Space Telescope has discovered thousands of extra solar planetary systems, proving that, far from being the exception, families of planets accompanying stars of every type are virtually the rule.[1] The just-launched Transiting Exoplanet Survey Satellite (TESS) and soon-to-be-launched Webb space telescope promise to be better still.[2] Webb will have nearly triple the diameter of Hubble and, stationed in an orbit around Earth-sun Lagrange point 2 (L2, on the far side of the Earth from the sun), will be able to explore the deepest recesses and deepest past of the universe in the infrared frequencies necessary to see high-redshift objects dating from the dawn of time. This will give us a new and powerful window, allowing us to see a vast cosmos invisible to either ground-based telescopes or Hubble. The planet-finding TESS telescope will expand Kepler's search for extrasolar worlds to more than four hundred times the area of sky that its pioneering predecessor was able to examine. While Kepler found thousands of previously unknown worlds, TESS could well find millions. Beyond these, NASA has plans for another generation of terrific space telescopes, able to study the universe with unprecedented power through windows blurred or completely blocked by the Earth's atmosphere. These include the WideField InfraRed Space Telescope (WFIRST),[3] the Gravitational Wave Surveyor,[4] the Cosmic Microwave Background Surveyor,[5] the Far InfraRed Surveyor,[6] the Lynx X-Ray Surveyor,[7] the Habitable Exoplanet Imaging Mission,[8] the Origins Space Telescope,[9] and, most important, the Large Ultra Violet Optical InfraRed (LUVOIR)[10] Surveyor. Currently planned for a circa-2030 launch to Earth-sun L2, LUVOIR (previously designated the Advanced Technology Large-Aperture Space Telescope, or ATLAST) will be an ultraviolet, optical, and near-infrared free-flying instrument whose 16-meter diameter will give it powers dwarfing both the 2.4-meter Hubble and the 6.5-meter Webb. Using these capabilities, LUVOIR will be able to resolve extrasolar planets, take the spectra of their atmospheres, and check them for free oxygen. Since free oxygen generally does not exist in nature in the absence of photosynthetic plants, detecting it would provide strong evidence for life among the stars.

The range of in-depth investigations we will be able to conduct with this set of instruments is incredible. We will be able to determine the frequency and number of potentially habitable planets in our galaxy. We will be able to detect seasonal variations and polar caps on such planets and learn much more about the process of planet formation itself. By resolving the moons of extrasolar planets, we will be able to determine their mass. We will be able to uncover the past and likely future history of the Milky Way, and many other galaxies, from birth to death. We will learn much more about the life history of stars, during both the distant cosmic past and today, and their role in the formation and circulation of the chemical elements throughout the galaxy. We will be able to hunt for black holes, from small to supermassive; study their collisions and growth; and investigate the inner working of quasars and other ultrahigh-energy ancient phenomena that provide extreme conditions to test our theories of physics. We will be able to measure the properties of neutron stars, thereby greatly expanding our insight into the nature of nuclear matter. We will be able to detect and study gravitational waves emitted from superenergetic events, including our own big bang (and perhaps others), subsequent inflation events, deflation events, big crunches, and all sorts of other happenings that may exist that we don't even have a clue about right now.

But we can go still further in our quest for knowledge. For example, we could build an array of optical telescopes on the moon.

It has been known for some time that the moon would be a superb location for astronomical observatories. One primary reason is that it has no obscuring atmosphere. A second is that it rotates only once every twenty-eight days, affording a telescope twenty-eight times as much time to gather light from a distant object as would be possible on Earth (or more than four hundred times as long as would be possible in Earth orbit). But in addition to being atmosphereless and slow turning, the moon is seismically dead and thus provides a rock-steady platform for mounting telescopes. This is an essential attribute required to create optical arrays in which groups of telescopes all focus on a single object and coordinate the signals they receive via computer. While the implementation of such arrays requires knowing the distance between telescopes to an accuracy of less than a millionth of a meter (and is thus nearly impossible to implement with substantial distances between telescopes on the seismically vibrating Earth or in free-floating space), their advantage to astronomy is extraordinary. The reason for this is that while the power of a single telescope to resolve detail is proportional to its diameter, the resolving power of an array of telescopes is proportional to the diameter of the array. So, while the Hubble Space Telescope has a diameter of 2.4 meters, an array of telescopes on the lunar surface could be stationed across the moon's diameter, some 3,400 kilometers, and achieve a resolution nearly a million times better than Hubble.

Such observatories would allow us to not only resolve but actually *map* Earthlike planets around nearby stars, perhaps out to a distance of a hundred light-years or so, within which some ten thousand solar systems are likely to be found. With the aid of the lunar optical array, we would be able to learn as much about these other solar systems as we knew about our own prior to the advent of interplanetary space probes in the 1960s. The array could also detect life—or possibly even civilizations—by taking spectra of planetary atmospheres and surfaces, checking them for free oxygen and chlorophyll as well as other biological and perhaps industrial or terraforming activity signatures.

Beyond this, the optical array would allow us to penetrate much deeper into the cosmos to examine and study in vastly greater detail objects dating from the time of the big bang that is generally believed to have initiated our universe.

Optical astronomy would not be the only branch of humanity's oldest science to benefit enormously from a lunar base. The moon's lack of atmosphere makes it an ideal platform for conducting cosmic ray, gamma ray, x-ray, and ultraviolet astronomy. These techniques, which are difficult to impossible to perform through Earth's thick atmosphere, are key to examining high-energy processes in the universe. The low temperatures available in the moon's permanently shadowed craters make them ideal locations for stationing infrared telescopes. The moon's far side is the only place in the solar system that is shielded from terrestrial civilization's massive radio chatter, so it is the best place for positioning radio telescopes. In addition, because the moon has no ionosphere, a radiotelescope positioned on the surface of the moon can pick up low-frequency (thirty megahertz or less) radio waves from space that the Earth's ionosphere completely masks from reception by ground-based instruments here. Each of these windows

254   THE CASE FOR SPACE

in the electromagnetic spectrum offers unique advantages and opportunities to discover new physical phenomena, and nearly all of them can best be studied from the moon.

This is of much more than academic interest. Historically, astronomy has led in the development of new physical laws. The laws of gravity, electromagnetism, relativity, and nuclear fusion were all discovered through astronomical observation. It is strange to think that the scientific basis for the technologies that dominate world commerce and power politics today, including accurate global navigation, ballistic missiles, electronic communications, and nuclear and thermonuclear weapons, all stem from the apparently impractical studies of astronomers. Yet this should be expected, since the universe provides a much bigger lab than any we can build. By allowing us to probe far deeper into time and space than ever before possible, a system of lunar and orbital observatories may well allow us to discover new laws of physics, including those of the creation process itself.

THE SEARCH FOR LIFE

But the moon is just the start. Exploring Mars will provide us the key to understanding the origin of life, which is, in fact, currently a mystery. Did life really originate on Earth? This idea is questionable because despite hundreds of years of searching by scientists, no free-living organisms simpler than bacteria have ever been discovered on our planet. This is truly remarkable; as simple as bacteria may be compared to more complex organisms, they are certainly not simple in any absolute sense, incorporating as they do, among other things, the entire elegant double-helix scripted language of DNA. Believing that bacteria were the first life-forms to emerge from chemistry is like believing that the iPhone was the first human-invented machine. This notion is clearly not credible. Just as the development of the iPhone had to be preceded by the development of computers, and radio, and telephones, and electricity, and glassware, and metallurgy, and written and spoken language, to name just a few necessary technological predecessors, so the creation of the first bacterium had to be preceded by the evolution of a raft of preceding biological technologies. But we see no evidence of any such history. We still see devices all around us that use one or more of the iPhone's ancestor technologies—telephones, light bulbs, batteries, glass windows, and steel knives, for example—but we see no prebacteria organisms. This observation has led many investigators, dating back to the great Swedish scientist Svante Arrhenius more than a century ago, to postulate that life on Earth is an immigrant phenomenon.[11] According to this "panspermia" hypothesis, bacteria did not originate on Earth but came here from space, after which they gave rise via generally understood evolutionary processes to all other life-forms.

But did microbes come from interstellar space, as Arrhenius argued? If so, we will find Earth-type bacteria on Mars, also without prebacteria, and know that the whole universe has been seeded with life, all of the same general biochemical and biological type. Or did life begin on our neighbor Mars? If so, we should be able to find not only bacteria but prebacterial organisms using more primitive versions of the same biochemistry. This would reveal life's past history, and thus the processes leading to its creation.

If, on the other hand, we find life utilizing a biochemistry differing from that of terrestrial organisms, that would support the idea of life originating separately on Earth, and almost certainly with great profusion and diversity everywhere else, as it would show that in two out of two planets, life appeared in two different ways almost immediately after it could. Furthermore, it would begin to reveal what features are essential to life and what are idiosyncratic to the Earth. All Earth life—from bacteria to mushrooms, oak trees, grasshoppers, crocodiles, raccoons, and humans—use the same biochemistry, based on the same twenty-one amino acids and the RNA/DNA system of encoding information. That's how we all do it, but is that what life is? Or are we simply one limited example drawn from a much vaster tapestry of possibilities? It is only by exploring other worlds that we can and will find out.

Finally, in the very unlikely event that we find no evidence of past or present life on Mars, this would tell us that despite the great similarity of early Mars to early Earth, life only appeared on our planet as a matter of chance, possibly freak chance, rather than lawful necessity. If so, we may very well be alone.

THE SEARCH FOR INTELLIGENCE

As for intelligent extraterrestrials, their communications and other activities such as interstellar transportation and terraforming that they might be engaged in are far more likely to be detected with the kinds of instruments or investigations that we can only locate in space.

For example, as noted, space telescopes could be used to detect oxygen in the atmospheres of extrasolar planets. By itself, this would simply indicate the presence of life. But what if it were found that while the typical frequency of such oxygenated planets is one per hundred solar systems, in one particular region of space, the frequency is one in five? That would strongly suggest that such a region was inhabited by an advanced spacefaring species engaged in active terraforming efforts.

Alternatively, one might seek evidence for the existence of advanced extraterrestrials by searching for the spectral signature emitted by their high-energy activities. The most obvious such activity to look for is interstellar travel.[12]

For example, an interstellar spaceship with a mass of one thousand tons wishing to accelerate or decelerate to or from a velocity of 10 percent of light speed within ten years would need to exert a power of about 1.5 terawatts. While there are a number of plausible methods to accelerate such a ship, the most advantageous way to decelerate it would be to use a strong magnetic field (i.e., a magnetic sail) to create drag against ionized or ionizable gas in the interstellar medium. Such a system would allow the ship to slow down, like a parachute behind a drag racer, without the need for propellant or power. (In fact, it would generate power.) In the process, it would also produce lots of long-wavelength radio waves, which would change in frequency as the ship slowed down in accord with some basic equations. As a result, it would be clearly identifiable as an artificial phenomenon. Unfortunately, however, the frequencies of such waves would be too low to penetrate the Earth's ionosphere. But they could be detected by suitable radio observatories in space.

The community involved in the Search for Extraterrestrial Intelligence (SETI) has for decades guided their efforts by the hypothesis that extraterrestrials would use submeter-wavelength radio for interstellar communication.[13] As noted above, the best place to conduct such research is the back side of the moon. However, it is a striking fact that despite searching since 1960 using ever more powerful instrumentation, ground-based SETI researchers have detected no such signals. One reason for this might be that the best electromagnetic frequencies for interstellar communication cannot penetrate the Earth's atmosphere. For example, if the ETs were using 0.1-micron-wavelength ultraviolet lasers for communication, instead of the 0.2-meter radio waves the SETI folks have been listening for, they could achieve two million times the data rate with a transmitter of the same size and power.

Such an increase in data rate would really come in handy, because even an advanced 0.03-meter-wavelength, hundred-watt X-band radio transmitter like the one on the Mars Reconnaissance Orbiter, which can deliver six megabytes per second broadcasting from Mars to a seventy-meter receiving dish on Earth, would only be able to deliver six microbits per second, or two hundred bits *per year*, if it were broadcasting from Tau Ceti some twelve light-years away. If we increased the transmitter power to a megawatt and its dish size from two to seventy meters, it would be able to deliver 0.6 bits per second. This still stinks, especially if one considers that twelve light-years is actually quite close as interstellar distances go.

So as convenient as such data transmission methods might be for the aspirations of ground-based SETI radio astronomers, it would be very impractical for ETs to use radio to talk over interstellar distances. If they are communicating between the stars, they would almost certainly be spacefaring themselves and thus make use of communication systems like ultraviolet lasers whose optimization is unconstrained by the need to penetrate planetary atmospheres. To hear such signals, we will have to listen for them *in space*.

But what if extraterrestrials are transmitting using entirely non-electromagnetic means? What if, instead of two-way dialogues, ETs are more interested in doing what we frequently have done when confronted with civilizations we find alien: spreading propaganda whose purpose is to urge others to be like us? (Think Radio Free Europe, or its Kremlin-sponsored adversary, Radio Moscow, both with the same fundamental message: "Be like us." Or the Gideons, leaving Bibles in every hotel room in the hope of converting travelers to their faith. Propaganda is propagation.) This sort of interstellar communication can be best done by broadcasting genetic packages.

COSMIC PROPAGANDA

The genetic material of individual common bacteria is estimated to contain between 130 kilobytes and fourteen megabytes of information. This material can be used as a powerful data storage medium, with a density of nine hundred terabytes per gram, about five hundred times the current state-of-the-art electromagnetic hardware. In experiments done to date, scientists have demonstrated such capabilities by encoding entire books and even movies in DNA and by showing that bacteria can be made to replicate encoded information when they reproduce.[14] In this way, microbes could be used to carry letters of the sort the SETI crowd have been looking for. ("Hi. Here's the value of pi. So now that you know we are intelligent and are paying attention to us, here's a dictionary of our language. And now that you've read that, here are the plans for building interstellar spaceships.")

But programmed bacteria could also be employed more practically to transmit genes, for the purpose of either establishing life on nonliving worlds or influencing the evolution of the biosphere of worlds already teeming with life. Bacteria are known to be able to transmit genes between species. That being the case, they could also be used to transmit genes between worlds.

The means for such transmission are available in every solar system, namely the light pressure exerted by the home star itself. Light has force, and sunlight has enough that engineers have designed solar sail spacecraft that, using vast sails made of thin reflective foils, could exploit its push to go sailing around the solar system.

Bacteria are so small and thin that if they were released individually into space, the light pressure on them from our sun would be greater than its gravitational attraction, and they would sail right out into interstellar space. Traveling at a typical velocity of 30 km/s (0.0001 c—the Earth's velocity around the sun), such microbes would take one hundred thousand years to fly ten light-years. This would expose them to cosmic ray doses between one and ten millirads, which is close to the limit for survivability of hardy microbial species such as radiodurans. This need not be a showstopper. A message sent using bacteria storage would no doubt use billions of individual microbes, and if even a few survived the trip, the message could still get through. While ultraviolet light would kill unshielded bacteria in days, effective shielding against this hazard can be provided by a half micron of soot. Such protection might conceivably occur naturally—thus raising the possibility of natural panspermia.[15] Regardless, such shielding could readily be provided by design in any artificial microsailcraft.[16]

For example, a sphere with a diameter of 4 microns surrounded by a disc with a diameter of 8 microns and a thickness of 1 micron would have the same surface/mass ratio as a simple sphere with a diameter of 1.2 microns and would be able to escape even a common, dim K star, let alone a brighter G star like our sun (which is bright enough to push a 4-micron-diameter sphere with a density equal to that of water to escape). Such microsailcraft designs could be readily mass-produced artificially. They could be stabilized in a manner to effectively serve as sails, either by spinning or by having inherently stable shapes, as exemplified by a badminton birdie. Upon arrival at a destination solar system, they would be decelerated by the light of its star and, because their sails provide them with such a large ratio of area to mass (i.e., a very low "ballistic coefficient"), safely enter planetary atmospheres and parachute down to the surface. They would then release their tiny passengers on new worlds, which, allowing them to multiply, would thereby act as natural receivers, amplifiers, and data storage devices for the biological information being transmitted.

Such craft could be projected across interstellar space at essentially no power cost to the transmitting party, beyond that required to launch them to planetary escape velocity. The means to implement such a program are actually well within our own capabilities today—and therefore would be clearly feasible for advanced extraterrestrials.

So, are they doing it? It would be very difficult to detect such microsailcraft on Earth, especially if they were designed, as they should be, to dissolve on contact with liquid water to release their payloads.

But in space, perhaps frozen on Titan, the Martian polar cap, comets, or Oort cloud objects, evidence of microsailcraft might conceivably be found and readily distinguished from anything native.

Scientists have discovered bacterial fossils known as stromatolites, dating back approximately 3.5 billion years, and residues of biological activities dating back 3.8 billion years—that is practically all the way back to the end of the heavy bombardment that previously made the early Earth uninhabitable. In fact, recently, a team of researchers reported microfossils that date back 4.28 billion years, *to the middle of the heavy bombardment*. In short, life appeared on our planet virtually as soon as it possibly could—and possibly several times, before it could last—strongly suggesting that it was already around, waiting to land and spread as soon as conditions on the ground were acceptable.

Was this an accident or part of a plan? Are we an accident, or part of a plan?

These are questions that thinking men and women have pondered for thousands of years. They are worth risking life and treasure to resolve. If we don't go, we won't know.

THE ULTIMATE MYSTERY

The ultimate mystery of the universe is not *what* it is but *why* it is. Why is the universe the way it is? Why does it exist at all?

There are those who argue that the second question above is beyond the scope of science. I'm not sure they are right. But clearly the first question is one that science needs to answer, and perhaps if it can, it can provide us with at least a clue to the second. So let's start with that.

The laws of physics can be written down in a number of equations providing the mathematical relationships between mass, charge, gravity, velocity, acceleration, distance, time, and so forth. While there are a priori reasons why the form of these equations should be of a certain type—for example, the gravitational force between two objects increasing in proportion to the product of their masses and decreasing in proportion to the square of the distance between their centers—the magnitude of the force is given by a constant that appears to be arbitrary.

In fact, physics is adorned with a whole raft of such constants, including the speed of light; Coulomb's constant; the Stefan-Boltzmann constant; Planck's constant; the gravitational constant; the permeability of free space; and the mass of the proton, electron, and other fundamental particles, among others, totaling some nineteen in all.

These nineteen constants could have had any value, and the laws of physics would still be mathematically consistent. However, it is easy to show that if even one of them were significantly different from its actual magnitude, life would be impossible.

In short, it seems that the laws of physics have been carefully fine-tuned in such a way as to make life possible.[17] How can we explain this?

One theory that has been advanced is known as the anthropic principle.[18] According to this theory, there is no mystery at all, because if the universe were not tuned for life, we would not be here to ask why it is. In my opinion, this is nonsensical. It is equivalent to answering the question "Why didn't the United States and the Soviet Union have a nuclear war?" with "Because if they did, you would be dead and not able to ask that question." In fact, one can explain anything the same way. Why did: it rain yesterday/the *Titanic* sink/the Union win the Civil War/the chicken cross the road? Because if it didn't, you wouldn't be asking why it did.

Equally useless is the multiverse theory, which states that all possible things happen and it's just that the alternatives to what we see occur in an infinite number of new universes that are constantly being created to accommodate them. Why did: it rain yesterday/the *Titanic* sink/the Union win the Civil War/the chicken cross the road? No reason. In other universes, it didn't.

The inadequacy of such pseudoscientific approaches has encouraged some who wish to offer religious explanations. Where there is design, there must be a designer, they argue. But, as proven by the success of the theory of evolution by natural selection, this assertion is false. Furthermore, the claim that supernatural intervention was responsible for any occurrence could just as easily be used to explain its nonoccurrence. It thus offers no explanatory power and represents little more than a cover for ignorance.

What is needed is a causal, naturalistic explanation for the fine-tuning of physical laws to make them friendly to life. A start at such a theory has been advanced by physicist Lee Smolin, who has written a very interesting book entitled *The Life of the Cosmos*.[19] Smolin suggests that universes are born within black holes, which are formed from stars, and that the universes born within black holes have similar, but not identical, laws to those of their parent stars. According to Smolin, if the daughter universes have better laws for producing stars (and thus black holes), they will multiply faster than those that produce few or no stars. Therefore, by a kind of natural selection, universes favoring stars, and thus life, would come to predominate, and our own life-friendly universe, far from being an unlikely anomaly, would be the odds-on favorite.

Smolin's theory can be regarded as quite speculative on any number of grounds. For example, it is unclear whether there are any other universes, and there seems little basis to either support or refute his notion that the laws of nature within a black-hole daughter universe should be "just a little bit different" from that of the parent universe.

But putting that aside, a larger problem is that, while necessary for life, stars could readily exist in universes having a much broader set of parameters than those which life requires. For example, life as we know it requires the existence of carbon, with all its peculiar characteristics that enable complex organic chemistry. Stars have no such requirement.

So, let me advance a hypothesis of my own, which is based on complexity theory. According to such theory, as set forth in other contexts by thinkers like Santa Fe Institute philosopher Stuart Kauffmann, order arises spontaneously in systems when A causes B and B causes A (or A causes B which causes C which causes D which causes A).[20] The free market works like this. Farmers grow crops that other people want to buy. That's why you can find food at the store. It's not a lucky accident without whose chance occurrence you would starve to death.

Ecosystems operate the same way. Plants are necessary for animals, which in turn play a vital role in spreading and fertilizing the seeds of plants. The causal relationship is clear. Natural systems spontaneously self-organize because if B is necessary for A, then for A to persist, it must and will do what is necessary to create and sustain B.

Following this logic, *if the physical universe has been finely tuned to allow for the development of life, it can only be because somehow life is necessary for the development of the physical universe.*

How can this be? Frankly, I don't know. It's quite clear that life has radically altered the chemistry and climate of the Earth to make it a much friendlier place for life, but changing the laws of physics is another matter altogether. Through its creation of intelligent forms capable of developing technology, life is clearly becoming more potent. We will soon have the power to terraform planets and travel between the stars. For the past two hundred years, the amount of energy humanity wields has increased at a rate of 3 percent per year, or tenfold per century. If we continue at this rate for another 1,200 years, we will be wielding power equal to the total output of the sun. Continue for 2,400 years—still a mere blink of an eye on the cosmic timescale—and our power will equal that of all the stars in the Milky Way galaxy combined. Let's say it takes ten, a hundred, or even a thousand times that long to achieve such potency—it hardly matters.

A species with anything like such powers and that understands the laws of creation might well be able to put its stamp on cosmic development—either by influencing the mechanism that Smolin has proposed or through other means entirely.

Could it be that it is not stars but intelligence that is responsible for the propagation and self-perfection of universes?

One thing is for sure: there's a lot that has happened, and still more happening, that we don't know about. I suggest we go up, look around, and start finding out.

FOCUS SECTION: MISTAKES IN THE DRAKE EQUATION

There are four hundred billion stars in our galaxy, and it's been around for ten billion years. Clearly it stands to reason that there must be extraterrestrial civilizations. We know this because the laws of nature that led to the development of life and intelligence on Earth must be the same as those prevailing elsewhere in the universe.

Hence, they are out there. The question is: how many?

In 1961, radio astronomer Frank Drake developed a pedagogy for analyzing the question of the frequency of extraterrestrial civilizations. According to Drake, in steady state, the rate at which new civilizations form should equal the rate at which they pass away, and therefore we can write:

$$\text{rate of demise} = N/L = R_* f_p n_e f_l f_i f_c = \text{rate of formation} \qquad (9.1)$$

Equation 9.1 is therefore known as the "Drake equation."[21] Here, N is the number of technological civilizations in our galaxy, and L is the average lifetime of a technological civilization. The left-hand term, N/L, is the rate at which such civilizations are disappearing from the galaxy. On the right-hand side, we have R_*, the rate of star formation in our galaxy; f_p, the fraction of these stars that have planetary systems; n_e, the mean number of planets in each system that have environments favorable to life; f_l, the fraction of these that actually developed life; f_i, the fraction of these that evolved intelligent species; and f_c, the fraction of intelligent species that developed sufficient technology for interstellar communication. (In other words, the Drake equation defines a "civilization" as a species possessing radiotelescopes. By this definition, civilization did not appear on Earth until the 1930s.)

By plugging in numbers, we can use the Drake equation to compute N. For example, if we estimate L as fifty thousand years (ten times recorded history), R_* as ten stars per year, f_p as 0.5, and each of the other four factors n_e, f_l, f_i, and f_c equal to 0.2, we calculate the total number of technological civilizations in our galaxy, N, equals four hundred.

Four hundred civilizations in our galaxy may seem like a lot, but scattered among the Milky Way's four hundred billion stars, they would represent a very tiny fraction: just one in a billion, to be precise. In our own region of the galaxy, (known) stars occur with a density of about 1 in every 320 cubic light-years. If the calculation in the previous paragraph were correct, it would therefore indicate that the nearest extraterrestrial civilization is likely to be about 4,300 light-years away.

But, classic as it may be, the Drake equation is patently incorrect. For example, the equation assumes that life, intelligence, and civilization can only evolve in a given solar system *once*. This is manifestly untrue. Stars evolve on timescales of billions of years and species over millions of years, and civilizations take mere thousands of years. Current human civilization could knock itself out with a thermonuclear war, but unless humanity drove itself into complete extinction, there is little doubt that a thousand years later, global civilization would be fully reestablished. An asteroidal impact on the scale of the K-T event that eliminated the dinosaurs might well wipe out humanity completely. But five million years after the K-T impact, the biosphere had fully recovered and was sporting the early Cenozoic's promising array of novel mammals, birds, and reptiles. Similarly, five million years after a K-T–class event drove humanity and most of the other land species to extinction, the world would be repopulated with new species, including probably many types of advanced mammals descended from current nocturnal or aquatic varieties. Human ancestors thirty million years ago were no more intelligent than otters. It is unlikely that the biosphere would require significantly longer than that to recreate our capabilities in a new species. This is much faster than the four billion years required by nature to produce a brand-new biosphere in a new solar system. Furthermore, the Drake equation also ignores the possibility that both life and civilization can propagate across interstellar space.

So, let's reconsider the question.

ESTIMATING THE GALACTIC POPULATION

There are four hundred billion stars in our galaxy, and about 10 percent of them are good G- and K-type stars that are not part of multiple stellar systems. Almost all of these probably have planets, and it's a fair guess that 10 percent of these planetary systems feature a world with an active biosphere, probably half of which have been living and evolving for as long as the Earth. That leaves us with two billion active, well-developed biospheres filled with complex plants and animals, capable of generating technological species on timescales of somewhere between ten million and forty million years. As a middle value, let's choose twenty million years as the "regeneration time" t_r. Then we have:

rate of demise = $N/L = n_s f_g f_b f_m / t_r = n_b / t_r$ = rate of creation (9.2)

where N and L are defined as in the Drake equation and n_s is the number of stars in the galaxy (four hundred billion), f_g is the fraction of them that are "good" (single G- and K-type) stars (about 0.1), f_b is the fraction of those with planets with active biospheres (we estimate 0.1), f_m is the fraction of those biospheres that are "mature" (estimate 0.5), and n_b, the product of these last four factors, is the number of active mature biospheres in the galaxy.

If we stick with our previous estimate that the lifetime, L, of an average technological civilization is fifty thousand years and plug in the rest of the above numbers, equation 9.2 says that there are probably *five million* technological civilizations active in the galaxy right now. That's *a lot* more than suggested by the Drake equation. Indeed, it indicates that one out of every eighty thousand stars warms the home world of a technological society. Given the local density of stars in our own region of the galaxy, this implies that the nearest *center* of extraterrestrial civilization could be expected at a distance of about 185 light-years.

Technological civilizations, if they last any time at all, will become starfaring. In our own case (and our own case is the only basis we have for most of these estimations), the gap between development of radio-telescopes and the achievement of interstellar flight is unlikely to span more than a couple of centuries, which is insignificant when measured against L at fifty thousand years. This suggests that once a civilization gets started, it's likely to spread. As we saw in chapter 7, propulsion systems capable of generating spacecraft velocities on the order of 5 percent the speed of light appear possible. However, interstellar colonists will probably target nearby stars, with further colonization efforts originating in the frontier stellar systems once civilization becomes sufficiently well-established there to launch such expeditions. In our own region of the galaxy, the typical distance between stars is five or six light-years. So, if we conservatively guess that it might take a thousand years to consolidate and develop a new solar system to the point where it is ready to launch missions of its own, this would suggest the speed at which a settlement wave spreads through the galaxy might be on the order of 0.5 percent the speed of light. However, the period of expansion of a civilization is not necessarily the same as the lifetime of the civilization; it can't be more, and it could be considerably less. If we assume that the expansion period might be half the lifetime, then the *average* rate of expansion, V, would be half the speed of the settlement wave, or 0.25 percent the speed of light.

As a civilization expands, its zone of settlement encompasses more and more stars. The density, d, of stars in our region of the galaxy is about 0.003 stars per cubic light-year, of which a fraction, f_g, of about 10 percent are likely to be viable potential homes for life and technological civilizations. Combining these considerations with equation 9.2, we can create a new equation to estimate C, the number of civilized solar systems in our galaxy, by multiplying the number of civilizations, N, by n_u, the average number of useful stars available to each.

$$C = Nn_u = (n_b L/t_r)(d)(f_g)(4\pi/3)(VL/2)^3 =$$
$$0.52(n_b/t_r)(df_g)V^3L^4 = 0.00016(n_b/t_r)V^3L^4 \qquad (9.3)$$

For example, we have assumed that the average life span, L, of a technological species is fifty thousand years, and if that is true, then the average age of one is half of this, or twenty-five thousand years. If a typical civilization has been spreading out at the above-estimated rate for this amount of time, the radius, R, of its settlement zone would be 62.5 light-years ($R = VL/2 = 62.5$ ly), and its domain would include about three thousand stars. If we multiply this domain size by the number of expected civilizations calculated above, we find that about 15 billion stars, or 3.75 percent of the galactic population, would be expected to lie within somebody's sphere of influence. If 10 percent of these stars are actually settled, this implies there are about 1.5 billion civilized stellar systems within our galaxy. Furthermore, we find that the nearest *outpost* of extraterrestrial civilization could be expected to be found at a distance of (185 − 62.5) = 122.5 light-years.

The above calculation represents my best guess as to the shape of things, but there's obviously a lot of uncertainty in the calculation. The biggest uncertainty revolves around the value of L; we have very little data to estimate this number, and the value we pick for it strongly influences the results of the calculation. The value of V is also rather uncertain, although less so than *L*, as engineering knowledge can provide some guide. (However, as hard science fiction writer/astrophysicist David Brin has forcefully argued, the rate of expansion *V* could be greatly accelerated without adding to the propulsion requirements at all, simply by taking longer leaps between stars.) In table 9.1, we see how the answers might change if we take alternative values for *L* and *V* while keeping the other assumptions we have adopted constant.

TABLE 9.1. THE NUMBER AND DISTRIBUTION OF GALACTIC CIVILIZATIONS

	V=0.005 c	V=0.0025 c	V=0.001 c
L=10,000 years			
N (# civilizations)	1 million	1 million	1 million
C (# civilized stars)	19.5 million	2.4 million	1 million
R (radius of domain)	25 ly	12.5 ly	5 ly
S (separation between civilizations)	316 ly	316 ly	316 ly
D (distance to nearest outpost)	291 ly	304	311 ly
F (fraction of stars within domains)	0.048%	0.006%	0.0025%
L=50,000 years			
N (# civilizations)	5 million	5 million	5 million
C (# civilized stars)	12 billion	1.5 billion	98 million
R (radius of domain)	125 ly	62.5 ly	25 ly
S (separation between civilizations)	185 ly	185 ly	185 ly
D (distance to nearest outpost)	60 ly	122.5 ly	160 ly
F (fraction of stars within domains)	30%	3.75%	0.245%
L=200,000 years			
N (# civilizations)	20 million	20 million	20 million
C (# civilized stars)	40 billion	40 billion	18 billion
R (radius of domain)	500 ly	250 ly	100 ly
S (separation between civilizations)	131 ly	131 ly	131 ly
D (distance to nearest outpost)	0 ly	0 ly	31 ly
F (fraction of stars within domains)	100%	100%	44%

In table 9.1, N is the number of technological civilizations in the galaxy (five million in the previous calculation), C is the number of stellar systems that some civilization has settled (1.5 billion, above), R is the radius of a typical domain (62.5 light-years), S is the separation distance between the centers of civilization (185 light-years), D is the probable distance to the nearest extraterrestrial outpost (122.5 light-years), and F is the fraction of the stars in the galaxy that are within someone's sphere of influence (3.75 percent).

Examining the numbers in table 9.1, we can see how the value of L completely dominates our picture of the galaxy. If L is "short" (ten thousand years or less), then interstellar civilizations are few and far between, and direct contact would almost never occur. If L is "medium" (about fifty thousand years), then the radius of domains is likely to be smaller than the distance between civilizations, but not much smaller, and so contact could be expected to happen occasionally (remember, L, V, and S are *averages*; particular civilizations in various localities could vary in their values for these quantities). If L is a long time (more than two hundred thousand years), then civilizations are closely packed, and contact should occur frequently. (These relationships between L and the density of civilizations apply in our region of the galaxy. In the core, stars are packed tighter, so smaller values of L are needed to produce the same "packing fraction," but the same general trends apply.)

Any way you slice it, one thing seems rather certain: There's plenty of them out there.

What are these civilizations like? What have they achieved?

It would be good to know.

Chapter 10

FOR THE CHALLENGE

We have come recently to boast of a global economy without thinking of its implications, of how unfortunate we are in finding it. It would be more cheering if news should come that by some freak of the solar system another world had swung gently into our orbit and moved so close that a bridge could be built over which people could pass to new continents untenanted and new seas uncharted. Would those eager immigrants repeat the process they followed when they had that opportunity, or would they redress the grievances of the old earth by a new bill of rights . . . ? The availability of such a new planet, at any rate, would prolong, if it did not save, a civilization based on dynamism, and in the prolongation the individual would again enjoy a spell of freedom. . . .

It would be very interesting to speculate on what the human imagination is going to do with a frontierless world where it must seek its inspiration in uniformity rather than variety, in sameness rather than contrast, in safety rather than peril, in probing the harmless nuances of the known rather than the thundering uncertainties of unknown seas or continents. The dreamers, the poets, and the philosophers are after all but instruments which make vocal and articulate the hopes and aspirations and the fears of a people.

The people are going to miss the frontier more than words can express. For four centuries they heard its call, listened to its promises, and bet their lives and fortunes on its outcome. It calls no more.

—Walter Prescott Webb, *The Great Frontier*, 1951

A bit more than 125 years ago, a young professor of history from the then-relatively-obscure University of Wisconsin got up to speak at the annual conference of the American Historical Association. Frederick Jackson Turner's talk was scheduled as the last one in the evening session. A long series of obscure papers preceded Turner's address, yet the majority of the conference participants stayed to hear him. Perhaps a rumor had gotten afoot that something important was about to be said. If so, it was correct, for in one bold sweep, Turner held forth with a brilliant insight. It was not legal theories, precedents, traditions, or national or racial stock that was the source of America's egalitarian democracy, individualism, and spirit of innovation. It was the existence of the frontier.

"To the frontier the American intellect owes its striking characteristics," Turner said,

> That coarseness of strength combined with acuteness and inquisitiveness; that practical, inventive turn of mind, quick to find expedients; that masterful grasp of material things, lacking in the artistic but powerful to effect great ends; that restless, nervous energy; that dominant individualism, working for good and evil, and withal that buoyancy and exuberance that comes from freedom—these are the traits of the frontier, or traits called out elsewhere because of the existence of the frontier.
>
> For a moment, at the frontier, the bonds of custom are broken and unrestraint is triumphant. There is no tabula rasa. The stubborn American environment is there with its imperious summons to accept its conditions; the inherited ways of doing things are also there; and yet, in spite of the environment, and in spite of custom, each frontier did indeed furnish a new opportunity, a gate of escape from the bondage of the past; and freshness, and confidence, and scorn of older society, impatience of its restraints and its ideas, and indifference to its lessons, have accompanied the frontier.
>
> What the Mediterranean Sea was to the Greeks, breaking the bonds of custom, offering new experiences, calling out new institutions and activities, that and more the ever-retreating frontier has been to the United States.

The Turner thesis was an intellectual bombshell, which within a few years created an entire school of historians who proceeded to demonstrate that not only American culture but the Western progressive humanist civilization that America generally represented in its most distilled form resulted from the great frontier of global settlement opened to Europe by the age of exploration.

Turner presented his paper in 1893. Just three years earlier, in 1890, the American frontier had been declared closed: the line of settlement that had always defined the furthermost existence of western expansion had actually met the line of settlement coming east from California. Today, a century later, we face the question that Turner himself posed—what if the frontier is gone? What happens to America and all it has stood for? Can a free, egalitarian, democratic, innovating society with a can-do spirit be preserved in the absence of room to grow?

Perhaps the question was premature in Turner's time, but not now. Currently we see around us an ever more apparent loss of vigor of American society: increasing fixity of the power structure and bureaucratization of all levels of society; impotence of political institutions to carry off great projects; the proliferation of regulations affecting all aspects of public, private, and commercial life; the spread of irrationalism; the banalization of popular culture; the loss of willingness by individuals to take risks, fend for themselves, or think for themselves; economic stagnation and decline; decreasing belief in the value of new people, whether through birth or immigration; the deceleration of the rate of technological innovation; declining faith in a positive future . . . Everywhere you look, the writing is on the wall.

Without a frontier from which to breathe life, the spirit that gave rise to the progressive humanistic culture that America has offered to the world for the past several centuries is fading. The issue is not just one of national loss—human progress needs a vanguard, and no replacement is in sight.

The creation of a new frontier thus presents itself as America's and humanity's greatest social need. Nothing is more important: apply what palliatives you will, but without a frontier to grow in, not only American society but the entire global civilization based upon Western values of humanism, reason, science, and progress will ultimately die.

I believe that humanity's new frontier can only be in space.

Why in space? Why not on Earth, under the oceans or in a remote region such as Antarctica?

It is true that settlements on or under the sea or in Antarctica are entirely possible, and their establishment would be much easier than that of Martian colonies. Nevertheless, the fact of the matter is that at this point in history, such terrestrial developments cannot meet an essential requirement for a frontier—to wit, they are insufficiently remote to allow for the full, free development of a new society. In this day and age, with modern terrestrial communication and transportation systems, no matter how remote or hostile the spot, on Earth, the cops are too close. If people are to have the dignity that comes with making their own world, they must be free of the old.

WHY HUMANITY NEEDS SPACE

> *Everything has tended to regenerate them; new laws, a new*
> *mode of living, a new social system; here they are become men.*
> —Jean de Crèvecoeur,
> *Letters from an American Farmer*, 1782

The essence of humanist society is that it values human beings—human life and human rights are held precious beyond price. Such notions have been for several thousand years the core philosophical values of Western civilization, dating back to the Greeks and the Judeo-Christian ideas of the divine nature of the human spirit. Yet they could never be implemented as a practical basis for the organization of society until the great explorers of the age of discovery threw open a New World in which the dormant seed of humanism contained within medieval Christendom could grow and blossom forth.

The problem with Christendom was that it was fixed—it was a play for which the script had been written and the leading roles both chosen and assigned. The problem was not that there were insufficient natural resources to go around—medieval Europe was not heavily populated, and there were plenty of forests and other wild areas—the problem was that all the resources were owned. A ruling class had been selected and a set of ruling institutions, ideas, and customs had been selected, and by the law of "Survival of the Firstest," none of these could be displaced. Furthermore, not only had the leading roles been chosen, but so had those of the supporting cast and chorus, and there were only so many parts to go around. If you wanted to keep your part, you had to keep your place, and there was no place for someone without a part.

The New World changed all that by supplying a place in which there were no established ruling institutions, an improvisational theater big enough to welcome all comers with no parts assigned. On such a stage, the players are not limited to the conventional role of actors—they become playwrights and directors as well. The unleashing of creative talent that such a novel situation allows is not only a great deal of fun for those lucky enough to be involved, it changes the opinion of the spectators as to the capabilities of actors in general. People who had no role in the old society could define their role in the new. People who did not "fit in" in the Old World could discover and demonstrate that, far from being worthless, they were invaluable in the new, whether they journeyed there or not.

The New World destroyed the basis of aristocracy and created the basis of democracy. It allowed the development of diversity by allowing escape from those institutions that imposed uniformity. It destroyed a closed intellectual world by importing unsanctioned data and experience. It allowed progress by escaping the hold of those institutions whose continued rule required continued stagnation, and it drove progress by defining a situation in which innovation to maximize the capabilities of the limited population available was desperately needed. It raised the dignity of workers by raising the price of labor and by demonstrating for all to see that human beings can be the creators of their world. In America, from colonial times through the nineteenth century when cities were rapidly being built, people understood that America was not a place one simply lived—it was a place one helped build. People were not simply inhabitants of their world. They were makers of it.

A TALE OF TWO WORLDS

Consider the probable fate of humanity in the twenty-first century under two conditions: with a space frontier and without it.

In the twenty-first century, without a space frontier, there is no question that human cultural diversity will decline severely. Already, in the late twentieth century, advanced communication and transportation technologies have eroded the healthy diversity of human cultures on Earth. As technology allows us to come closer together, so we come to be more alike. Finding a McDonald's in Beijing, country and western music in Tokyo, or a Taylor Swift T-shirt on the back of an Amazon native is no longer a great surprise.

Bringing together diverse cultures can be healthy, as it sometimes results in fusions that produce temporary flowerings in the arts. It can also result in very unpleasant increases in ethnic tensions. But however the energy released in the cultural merger is expended in the short term, the important thing in the long term is that it is expended. An analogy to cultural homogenization is that of connecting a wire between the terminals of a battery. A lot of heat can be generated for a while, but when all the potentials have been leveled, a condition of maximum entropy is reached, and the battery is dead. The classic example of such a phenomenon in human history is the Roman Empire. The golden age produced by unification is frequently followed by stagnation and decline.

The tendency toward cultural homogenization on Earth can only accelerate in the twenty-first century. Furthermore, because of rapid communication and transportation technologies "shorting out" potential terrestrial intercultural barriers, it will become increasingly impossible to obtain the degree of separation required to develop new and different cultures on Earth. If the space frontier is opened, however, this same process of technological advance will also enable us to establish a new, distinct, and dynamic branch of human culture on Mars and eventually more on worlds beyond. The precious diversity of humanity can thus be preserved on a broader field, but only on a broader field. One world will be just too small a domain to allow the preservation and continued generation of the diversity needed not just to keep life interesting but to assure the survival of the human race.

Without the opening of a new frontier in space, continued Western civilization faces the risk of technological stagnation. To some, this may appear to be an outrageous statement, as the present age is frequently cited as one of technological wonders. In fact, however, the rate of progress within our society has been decreasing, and at an alarming rate. To see this, it is only necessary to step back and compare the changes that have occurred in the past forty years with those that occurred in the preceding forty years and the forty years before that.

Between 1900 and 1940, the world was revolutionized: cities were electrified, telephones and broadcast radio became common, talking motion pictures appeared, automobiles became practical, and aviation progressed from the Wright Flyer to the DC-3 and Hawker Hurricane. Between 1940 and 1980, the world changed again, with the introduction of communication satellites and interplanetary spacecraft; computers; television; antibiotics; nuclear power; Atlas, Titan, and Saturn rockets; Boeing 747s and SR-71s. Compared to these changes, the technological innovation from 1980 to the present seems insignificant. Immense changes should have occurred during this period but did not. Had we been following the previous eighty years' technological trajectory, we today would have flying cars, maglev (magnetic levitation) trains, fusion reactors, hypersonic intercontinental travel, reliable and inexpensive transportation to Earth orbit, undersea cities, open-sea mariculture, and human settlements on the moon and Mars. It is true that there has been noteworthy progress in the area of information technology, but the overall rate of transformation does not compare to that of the previous two periods. Instead, today we see important technological developments, such as nuclear power and biotechnology, being blocked or enmeshed in political controversy. The space launch revolution offers an opportunity to lift ourselves out of the thickening quagmire, but otherwise, overall, we are slowing down.

Now, consider a nascent Martian civilization: Its future will depend critically upon the progress of science and technology. Just as the inventions produced by the "Yankee ingenuity" of frontier America were a powerful driving force on worldwide human progress in the nineteenth century, so the "Martian ingenuity" born in a culture that puts the utmost premium on intelligence, practical education, and the determination required to make real contributions will make much more than its fair share of the scientific and technological breakthroughs that will dramatically advance the human condition in the twenty-first.

Prime examples of the Martian frontier driving new technology will undoubtedly be found in the areas of robotics and artificial intelligence (because of a shortage of available labor and diversity of skills), high-productivity agriculture (because of the limited land made available by greenhouse farming), and especially of energy production. As on Earth, an ample supply of energy will be crucial to the success of Mars settlements. The Red Planet does have one major energy resource that we currently know about: deuterium, which can be used as the fuel in nearly waste-free thermonuclear fusion reactors. Earth has large amounts of deuterium too, but with all of the existing investments in other, more polluting forms of energy production, the research that would make possible practical fusion power reactors has been allowed to stagnate. The Martian colonists are certain to be much more determined to get fusion online, and in doing so, they will massively benefit the mother planet as well.

The parallel between the Martian frontier and that of nineteenth-century America as technology drivers is, if anything, vastly understated. America drove technological progress in the last century because its western frontier created a perpetual labor shortage back East, thus forcing the development of labor-saving machinery and providing a strong incentive for improvement of public education so that the skills of the limited labor force available could be maximized. This condition no longer holds true in America. In fact, far from prizing each additional citizen, anti-immigrant attitudes are on the rise, and a vast "service sector" of bureaucrats and menials is being created to absorb the energies of those parts of the population whose participation in the productive parts of the economy is no longer needed. Thus, in the early twenty-first century, each additional citizen is increasingly regarded as a burden.

On late twenty-first-century Mars, on the other hand, conditions of labor shortage will apply with a vengeance. Indeed, it can be safely said that no commodity on Mars will be more precious, more highly valued, and more dearly paid for than human labor time. So rather than restricting immigration or suppressing human potential by imposing onerous licensing requirements to block entry into professions, Mars, like nineteenth-century America, will welcome immigrants with open arms, let everyone employ their talent any way they can, and do everything possible to multiply their productivity through technology. Workers on Mars will be paid more and treated better than their counterparts on Earth. Just as the example of nineteenth-century America changed the way the common person was regarded and treated in Europe, so the impact of progressive Martian social conditions may be felt on Earth as well as on Mars. A new standard may be set for a higher form of humanist civilization on Mars, and, viewing it from afar, the citizens of Earth will rightly demand nothing less for themselves.

The frontier drove the development of democracy in America by creating a self-reliant population that insisted on the right to self-government. It is doubtful that democracy can persist without such people. True, the trappings of democracy exist in abundance in America today, but meaningful public participation in the process is deeply wanting. Consider that no representative of a new political party has been elected president of the United States since 1860. Likewise, neighborhood political clubs and ward structures that once allowed citizen participation in party deliberations have vanished. This has led to a banalization of political culture, and popular culture generally, making the body politic increasingly vulnerable to demagogues. And regardless of the will of Congress, the real laws, covering ever-broader areas of economic and social life, are increasingly being made by a plethora of regulatory agencies whose officials do not even pretend to have been elected by anyone.

Democracy in America and elsewhere in Western civilization needs a shot in the arm. That boost can only come from the example of a frontier people whose civilization incorporates the ethos that breathed the spirit into democracy in America in the first place. As Americans showed Europe in the last century, so in the next the Martians can show us the path away from oligarchy.

It is quite simple: Societies and individuals share a characteristic. We grow when we accept challenge. We stagnate and decline when we do not. We see this pattern in human history again and again. Those societies that have achieved unchallenged dominance over their relevant domain have tended to crystallize into self-satisfied static forms, with some classic examples being ancient Egypt and traditional China. "We are the world; we have everything there is to have, we know everything there is to know, we have done everything there is to do" is the proud slogan of such fundamentally dead cultures. In contrast, those societies that have been open to accepting new challenges have proven the most dynamic.

It is true that in the past, war has also sometimes imposed challenges on societies, which forced those that rose to meet them to grow against their inclination (while eliminating those that didn't). But warfare is destructive of both the wealth and human potential of a society, and as the level of technology advances, so does the level of destruction. With the advent of nuclear weapons and recombinant-DNA-based bacteriology, warfare of a sufficiently serious nature to induce societal stress among leading states has become unthinkable, as it would lead to the collapse of civilization itself.

Frontier shock, the stress induced in a people when they take on the challenge of strange new lands filled with new possibilities and new knowledge, has been a much more interesting and dynamic force. This has been true ever since our species began.

In a biological sense, humans are not really native to the Earth. We are tropical creatures, native to Kenya. That is why we have long, thin arms with no fur. Across most of this planet, unprotected human life for any length of time is as impossible as it is on the moon. We survive and thrive outside our African natural habitat solely by virtue of our technology.

The move outward from our birthplace did not occur quickly. For 150,000 years after the appearance of *Homo sapiens*, our ancestors remained in the tropics. With the aid of a few simple, crude stone implements inherited from their *Homo erectus* forebears, these early *Homo sapiens* were masters of their environment and apparently saw little need to either move or change in any way. Indeed, the 150 millennia humanity spent in Africa were a period of almost total technological stagnation, with generation after generation after generation living and dying doing things in exactly the same way as their parents, grandparents, and remote ancestors centuries, millennia, and tens of millennia before.[1]

But then, for some reason, about fifty thousand years ago, some bands of these people left the African homeland to try their fortunes in the north. There, they soon encountered the problems of life in the wintry wastes of Ice Age Europe and Asia. In this new and more challenging world, the old bag of tricks that had served static tropical humans so well for so long no longer sufficed. Indeed, without the novel inventions of clothing, insulated shelter, and efficient control of fire, *Homo sapiens* could not survive a single winter in their new habitat. Inventing clothing meant inventing sewing. Shelters had to be either built de novo or won from powerfully built, stocky, cold-adapted Neanderthals or 1,500-kilogram cave bears. Moreover, these wanderers were no longer in a world where food could be reliably gathered all year long. Dealing with these challenges required fine-tooled weapons that could kill at a distance for combat and big game hunting, and improved means of communication, planning, and coordination among *Homo sapiens* themselves. Thus, we were forced to develop language and other forms of symbolic communication. Within a few thousand years of their arrival in the north, we find our ancestors making all sorts of novel gear, including a wide array of finely chipped and polished stone tools and weapons as well as bone tools that included sewing kits and fishing kits, and producing fine cave art and even musical instruments. The latter two innovations are especially significant. Many animals build shelters, and sea otters, chimpanzees, and crows have all been known to use simple tools. But creating symbolic art—that's something else. Of all the creatures of this Earth, only humans *paint*. The rendering and appreciation of visual images denotes a mental ability akin to that required to create and understand verbal images. In other words, it indicates the origin of language, and with it, in all probability, stories, mythology, oral history, poetry, and songs.[2] A qualitatively higher level of intellectual—and, I would argue, spiritual—development had been attained.

Moving into a more challenging environment to which it was not naturally adapted, inspired *Homo sapiens* to transcend itself. Instead of existing as a clever animal applying a fixed repertoire of abilities to deal with a fixed set of contingencies in a well-defined environment, we became a species whose fundamental means of dealing with the world is to constantly invent new abilities. *Homo sapiens* became

Homo technologicus, man the inventor, and by so doing enabled itself to conquer all the environments of the world: deserts, forests, jungles, steppes, swamps, mountains, tundra, rivers, lakes, seas, and oceans.

Then, throughout recorded human history, the most progressive cultures have been the "Sea People," such as the Minoans, the Phoenicians, the Greeks, diaspora Jews, Italian Renaissance city-states, the Hanseatic League, the Dutch, the British, and the Americans, whose leading elements constantly accepted new challenges by engaging in long-range (typically maritime) trade and/or exploration. Societies of "Land People" whose top elements have been drawn from a landed aristocracy ruling a fixed domain have had a much more limited view and thus generally been far more tradition bound and conservative. The greatest stimulus of all occurs in those situations where not just a leading minority but large fractions of a society's population are exposed to or immersed in the novel frontier environment, where they are both forced and free to innovate. Thus, it is no coincidence that the blossoming of classical Greek culture occurred during and immediately following their age of Mediterranean colonization, or that the fantastic explosion of innovation in European culture that transformed unimpressive and relatively static medieval Christendom into hyperdynamic and globally dominant Western civilization occurred simultaneously with the West's age of discovery and colonization. The most extreme example of the stimulus of frontier shock is North American civilization, which was developed as a culture of innovation, antitraditionalism, optimism, individualism, and freedom based on four hundred years of formative interaction with the novel necessities and infinite possibilities posed by its vast and constantly changing frontier.[3]

But what of now? *Pax Terrestris*, yes. *Pax Mundana*, no. Humanity does not need war, death, disease, decay, superstition, national or racial cults, archaic belief structures or despotisms, or any number of other residues of our primitive past against which many noble people have struggled through the ages. But humanity does need challenge. A humanity without challenge would be a humanity without change, without innovation, which fundamentally means a humanity without meaningful freedom. A humanity without challenge would be a humanity without humanity.

Furthermore, the "golden age" enjoyed by a static society is generally only a transitory phase on the path to hell. This is so because, like the hardscrabble poor kid who, after succeeding in life, ensures through his determination that his children "should never have to struggle the way he did" that they never develop the toughness that allowed him to make his fortune, the balmy conditions of the golden age tend to undermine the character of the people that allowed them to achieve it. Thus, in his seminal work on world history, *The Evolution of Civilizations*, historian Carroll Quigley identified seven major stages in the development of societies: (1) mixture, (2) gestation, (3) expansion, (4) conflict, (5) universal empire, (6) decay, and (7) collapse.[4] With its victory in the Cold War circa 1990, Western (essentially modern global) civilization reached stage five. Should we choose to continue in the footsteps of such historical analogs, stage six would soon follow—and in fact, some would argue that it has already begun.

In 1992, philosophy professor Francis Fukuyama wrote a widely read book entitled *The End of History*, in which he posited that with the unification of the world resulting from the West's victory in the Cold War, human history had essentially "ended."[5] In 1996, *Scientific American* writer James Horgan published a much more interesting best seller entitled *The End of Science*, in which he held that all the really big discoveries to be made in science had already been made, and thus the enterprise of scientific discovery must soon grind to a halt.[6] (The day after I finished reading Horgan's book in February 1998, a group of astronomers using the Hubble Space Telescope announced they had found a fifth fundamental force in nature.) In his book, Horgan interviewed Fukuyama and asked him what he thought of those who doubt we have reached the end of human history. "They must be space travel buffs," Fukuyama replied in derision. Indeed.

The Earth's challenges have largely been met, and the planet is currently in the process of effective unification. I believe this marks the end, not of human history but merely of the *first phase* of human history: our development into a mature Type I civilization. It is not the end of history, because if we choose to embrace it, we have in space a new frontier offering endless challenge—an infinite frontier, filled with worlds waiting to be discovered and history waiting to be made by myriad new branches of human civilization waiting to be born.

Are we living at the end of history or at the beginning of history? Are we old, or are we young?

The choice is ours.

FOCUS SECTION: SPACE PROGRAM SPIN-OFFS

One of the main selling points that NASA has frequently advanced to support its funding are the technological advances developed to meet space program needs that have greatly benefited society at large. These "spin-off" claims have sometimes drawn derision from skeptics because many, such as Tang, are trivial, and several of the most often cited, including Velcro, bar codes, and Teflon, were actually invented in the 1940s or early 1950s, well before NASA was created.

Nevertheless, the fact of the matter is that by accepting the harsh challenges of space missions, NASA-sponsored R&D efforts have actually produced a very impressive array of spin-off technologies.

Among the spin-offs are thousands of inventions or major improvements in the fields of computer technology, environment and agriculture, health and medicine, public safety, transportation, recreation, and industrial productivity. These include MRI scanning, infrared ear thermometers, ventricular assist devices, artificial limbs, light-emitting diodes in medical therapies, surgical technologies, invisible braces, scratch-resistant lenses, emergency blankets, aircraft anti-icing systems, highway safety devices, chemical detection systems, radial tires, video enhancing and analysis systems, fire-resistant materials, firefighting equipment, smoke detectors, air-conditioning systems, temper foam, enriched baby food, freeze-dried food, portable cordless vacuum cleaners, cordless power tools, digital image sensors, quartz clocks, water purification systems, solar cells, pollution remediation systems, microchips, structural analysis software, remotely controlled ovens, powdered lubricants, mine safety devices, food safety systems, and 3-D printing systems.

Many thousands of lives have been saved by NASA-funded medical, firefighting, and safety technologies. Virtually nothing of consequence is built anymore without use of NASA-invented design software packages such as NASTRAN, and NASA contributions to advancing computer technology are being felt in all areas of our economy. Efficient silicon solar cells were first developed at NASA's instigation, and if solar power ever becomes a major source of energy for our society, humanity will owe a lot of thanks to NASA for the gift.

In addition to these terrestrial spin-offs, NASA technology used in space has provided profound benefits in weather prediction, communications, and navigation, with massive impacts on the entire economy.

So it really is true: the space program has been a major driver of technological progress, yielding innovations whose return to society has greatly exceeded the cost of investment.

But the greatest value we have received from accepting the challenge of space exploration has come not in the form of particular inventions but from its production of human intellectual capital.

This can be seen by looking at how young people have responded to the existence of a bold space program—or lack of it—in making their life decisions. Figure 10.1 shows the number of college degrees granted in the United States in the STEM fields of science, engineering, and mathematics at every level in the 1960s, 1970s, and 1980s.[7]

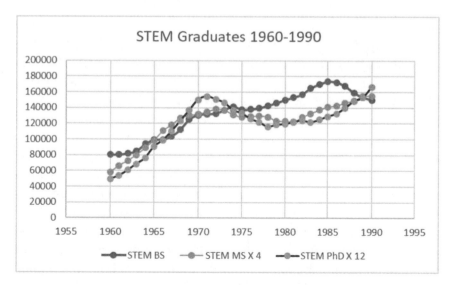

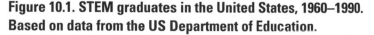

Figure 10.1. STEM graduates in the United States, 1960–1990. Based on data from the US Department of Education.

It can be seen that in the Apollo-era years, from 1960 to 1972, the number of STEM graduates with bachelor's degrees increased 70 percent, those with master's degrees rose 130 percent, while those with doctorates soared by a whopping 240 percent! However, virtually immediately following the end of the Apollo program in 1972, these increases halted, with numbers of advanced degrees actually going into immediate decline in absolute terms, and bachelor's degrees stagnating in absolute terms while declining relative to population size.

Today, American education is in decline, which members of the political class are attempting to remedy by the counterproductive expedient of massive standardized testing programs that are only serving to waste class time and school systems' money while deepening students' adverse attitudes toward education. But youth loves adventure, and a societal commitment to opening the space frontier would be an invitation to adventure to every boy and girl in every country taking part, calling out the stirring challenge: "Learn your science, and you can be an explorer of new worlds." During the Apollo period, the number of science graduates in the United States increased radically at every level—high school, college, and PhD. We have benefited from that intellectual capital ever since. A humans-to-Mars program today would have even greater impact—since the scientific professions are now open to women and minorities in a way that was simply not the case in the 1960s—producing millions of young scientists, engineers, inventors, doctors, medical researchers, and technological entrepreneurs whose innovations would return benefits to society exceeding the costs of the space effort hundreds of times over.

The fact that the United States has, with all its limitations, the best space program in the world has been a major factor in drawing young people of talent to America. In comparison, the European Space Agency, for example, despite being supported by a larger population and economy than the United States, has one-fifth NASA's budget, and Australia has virtually no space agency at all. Effectively, the leaders of such countries have told their young scientists; "if you want to take part in space exploration, emigrate." That message has been heard, loud and clear, and the resulting brain drain has been a tremendous loss to all such badly led nations.

Fortune favors the bold.

Chapter 11

FOR OUR SURVIVAL

All civilizations become either spacefaring or extinct.
—Carl Sagan[1]

IN THE LINE OF FIRE

On October 1, 1990, US strategic defense forces detected an explosion in the Central Pacific, with a power equal to the ten-kiloton atomic bomb that leveled Nagasaki at the end of World War II. The French government, it is known, sometimes tests nuclear weapons in the Pacific, but this was not one of theirs. No earthly power was behind it. This missile came from outer space. Exploding in the air above the uninhabited central Pacific, it caused no known casualties. However, had it arrived some ten hours later, it would have detonated in the Middle East, right in the midst of a powder keg loaded with a vast United Nations coalition and Iraqi armed forces then preparing for all-out war.

The October 1, 1990, projectile came from the enemy's lightest artillery, and it flew wide of any vital target. But there's a lot more where that came from, and we've been hit by far worse. And not so long ago.

For example, on February 12, 1947, a larger extraterrestrial projectile with a hundred-kiloton explosive force (ten times the Nagasaki bomb) impacted less than four hundred kilometers from the Russian city of Vladivostok. This may have caused casualties among the large labor camp population that adorned the region at that time. Records of the missing, and the reactions of the no doubt bewildered authorities, may perhaps still be found in musty Stalinist archives.

A much larger hit occurred on June 30, 1908, when an extraterrestrial warhead with a force of twenty thousand kilotons (twenty megatons!) hit Tunguska, Siberia, leveling thousands of square kilometers of forest. Thousands of deer were killed, and probably hundreds of Siberian hunters whose lives were beneath the notice of the Tsarist regime as well. Had the object arrived only three hours later, it would have devastated a large area of European Russia, and possibly caused the cancellation of one of the social events at the Tsar's summer palace.

The most recent hit to make the news occurred February 15, 2013, when a previously undetected microasteroid twenty meters in diameter entered the Earth's atmosphere at a velocity of nineteen kilometers per second and exploded at thirty kilometers altitude near the Ural mountain city of Chelyabinsk. The 450-kiloton explosion (thirty times the yield of the Hiroshima bomb) generated a bright flash and a powerful shock wave, injuring 1,500 people and damaging 7,200 buildings. The event took place on the same day that the larger asteroid 367943 Duende made its well-predicted and well-publicized close pass by Earth. According to the authorities who were caught by surprise by the Chelyabinsk meteor, it was an unrelated event. Believe them if you like.

These well-documented impacts all occurred within living memory. Similar events doubtlessly occurred at a comparable rate in preceding centuries, but due to the prevalence of mysticism, illiteracy, lack of scientific outlook, and poor global communications in prior centuries, humanity's capability to reliably observe, communicate, and record earlier impacts was very weak. Perhaps some of the cataclysmic events described in ancient religious and mythological texts, such as the destruction of Sodom and Gomorrah, are folkloric accounts of asteroid impacts.[2] It's hard to say; but since both asteroids and the true origin of meteorites were unknown until the nineteenth century, an earlier impact event could probably only be explained by its surviving observers as a manifestation of the rage of God.[3]

But if premodern humanity was an unreliable recorder of asteroid bombardment, there was another scribe on the scene, whose accounts, while incomplete, never lie—the Earth itself. The geological record shows without room for doubt that the Earth has been the target for massive asteroid bombardment throughout its history (see map in figure 11.1).[4]

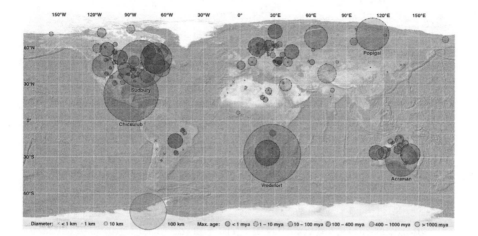

Figure 11.1. Known asteroid impact craters are scattered all over the Earth. Size of circles depicts relative energy of impact. *Image courtesy of Impact Field Studies Group.*

Of course, to make it into the Earth's record, an asteroid hit has to be something *significant*, not a mere A-bomb- or H-bomb-sized blast like October 1990 or Tunguska. For example, the kilometer-wide, two-hundred-meter-deep crater in northern Arizona visited by many tourists today is the scar left by the impact of a 150-meter-diameter asteroid that hit the Earth fifty thousand years ago with a force of about 840 megatons (eighty-four thousand Nagasaki bombs), releasing roughly the same explosive power as would have been detonated had an all-out nuclear war occurred between NATO and the Warsaw Pact at the most heavily armed phase of their balance of terror. The impact probably wiped out much of the life in the American Southwest and raised enough dust to throw the entire world into a "nuclear winter" deep freeze for several years.

But the 150-meter rock that hit Arizona was still a *small* asteroid. There are others out there with masses thousands, even *millions* of times as great. Sometimes they hit too. Table 11.1 provides a rough guide to how much force is released when that happens and the estimated frequency of such events.

TABLE 11.1. DESTRUCTIVE FORCE OF ASTEROID IMPACTS

Asteroid diameter	Explosive yield (kT)	Frequency	Example
4 m	16	1/year	Central Pacific, Oct. 1, 1990
8 m	128	1/decade	Vladivostok, Feb. 12, 1947
43 m	20,000	1/400 years	Tunguska, June 30, 1908
150 m	844,000	1/6,000 years	Arizona, 50,000 BCE
1 km	240 million	1/300,000 years	South Pacific, 2.3 million BCE
10 km	240 billion	1/60,000,000 years	Yucatan, 65 million BCE

(See plate 16.)

The asteroid impact of 65 million BCE is the most famous of the large events, as its discovery has led to a revolution in our understanding of the history of life on Earth. Prior to the 1980s, scientists believed that the evolution of species followed a gradual course, driven only by interaction among species and between life and the Earth's climate. Then, in 1980, the team of Nobel Prize–winning physicist Luis Alvarez and his son, paleontologist Walter Alvarez, discovered a thin layer of iridium in Italian sediments located precisely at the sixty-five-million-year-old boundary between the Cretaceous period, in which dinosaurs dominated the Earth, and the Tertiary period, in which dinosaurs no longer existed.[5] Iridium is rare on Earth but common in meteorites, which represent samples of small asteroids. The sudden disappearance of the long-lasting, widely distributed, and widely varied order of dinosaurs had always been a mystery in paleontology. Discovering asteroid material deposited precisely at the time of the dinosaur's demise in Italy and then all around the world, the Alvarezes advanced the startling theory that an asteroid impact had wiped the giant reptiles out. Traditional paleontologists wedded to older ideas resisted the hypothesis, but the Alvarezes' case became conclusive in 1991 when Canadian geologist Alan Hildebrand discovered the crater left by the dinosaurs' destroyer at Chicxulub in the Yucatan Peninsula of Mexico.

The size of the crater reveals the size of the impact. With an explosive force of more than two hundred billion kilotons of high explosive, the question is no longer what killed the dinosaurs. The real mystery is how anything else survived. First, the hypersonic shock created when the asteroid tore into the atmosphere turned the air to superhot plasma, instantly baking anything within line of sight of the entry trajectory. Then the asteroid slammed into the Earth itself, shooting vast amounts of ejecta into space, which later reentered at hypersonic speed and heated the entire atmosphere to incandescence. The glowing sky set fire to forests everywhere. Any life that could not find a hiding place underground or underwater was killed. The burning only lasted a few days, but afterward the dust released by the impact and consequent smoky fires caused intense, planetwide lethal acid rain and sent the Earth into years or decades of a dust-shrouded deep nuclear winter. Eventually the dust rained out, allowing the blessed sunlight to warm the Earth again, except that the impact and the fires had put a massive amount of carbon dioxide gas into the atmosphere, which the sparse vegetation of the post-holocaust world could do little to clean up. As a result, a powerful greenhouse effect was created, rapidly driving the Earth into an intolerable hothouse that may have lasted for centuries.

Subsequent research has revealed that the dinosaurs' killer was not unique and that many, if not all, of the dozens of mass extinctions that define the successive ages of the Earth's paleontological history were caused by asteroid impacts. The body count of these events is very high: during *each* of them, 35–95 percent of all living species of plants and animals on the planet were exterminated. More than two-thirds of all species that have ever lived on Earth were wiped out by asteroids.

So be proud of your ancestors, who survived all this. They were *very* tough. But it seems unwise to keep pushing our luck. Look at figure 11.2, which shows the orbits of known Earth-crossing asteroids. About two hundred are known, but it is estimated that there are at least two thousand of them with diameters greater than one kilometer, and two hundred thousand bigger than one hundred meters. About 20 percent of them will hit the Earth sooner or later. In most cases, later means not for hundreds of millions of years. But not all. And somewhere out there, right now, there is an asteroid with your descendants' names written on it. It's not on a direct collision course; it will orbit the sun many times before it hits. But as complex as its path might be, if nothing is done, the laws of gravity and celestial mechanics dictate that it will unerringly follow its prescribed course to bring death to your progeny and myriads of other species who passively await its coming.

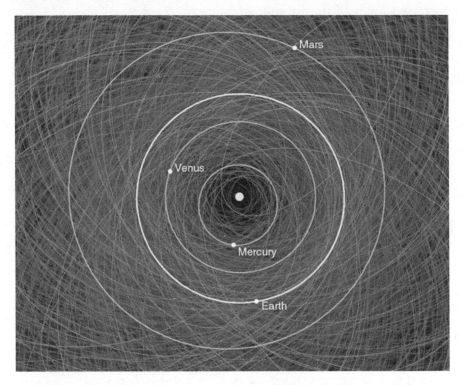

Figure 11.2. Asteroid trajectories. The Earth is in the line of fire. *Image courtesy of NASA Jet Propulsion Laboratory.*

We are targets in a cosmic shooting gallery. Prudence dictates that we take steps to rectify this condition.

MOVING ASTEROIDS

The good news is that all we have to worry about is a couple of thousand modest asteroids with explosive yields of a few hundred million kilotons of dynamite each, plus maybe a few million more tiny fellows with yields of tens or hundreds of thousands of kilotons. Possibly because political office offers little protection against asteroid impacts, Congress has found this threat sufficiently compelling to spend a few million dollars on a NASA Near-Earth Object Observation Program (NEOOP) to gradually identify and begin to track potential doomsday projectiles. Equipped only with a few smallish or obsolescent telescopes, NEOOP is gradually charting the existing major hazards. Presumably we've got the time. Smaller, hundred-meter objects capable of delivering the punch of several hundred H-bombs, will, however, not be detected. This seems to me to be a significant shortcoming in the program because such objects hit us much more frequently than the planet killers. Also, if we could detect them, it would be much more feasible to take positive action to deflect one of the small guys. Which brings us to the $64,000 question: If we do detect a world-destroying asteroid coming our way, what do we do about it?

There are a number of conceivable alternatives. These include:

(a) Sit tight and die. This is the traditional approach, which wants improvement.
(b) Evacuate the Earth. This will be technically infeasible for some time to come, and always undesirable.
(c) Move the Earth out of the way. Amusing but technically infeasible.
(d) Destroy the asteroid prior to impact. This won't work. It's probably infeasible to fragment a one-kilometer asteroid with weaponry, but even if we could, it would do little good. The fragments hitting the Earth would do nearly as much damage as the asteroid would have if left in one piece.
(e) Deflect the asteroid so that it misses the Earth. This would be difficult but is possible in principle. It has thus justly received the most attention by those concerned with planetary defense.

How can we deflect an oncoming asteroid? This was the subject of a workshop held at Los Alamos National Lab in 1992 and a number of similar locales since. Most of the technical contributors to these meetings are designers of advanced thermonuclear weapons from Los Alamos and Livermore, so naturally the focus generally falls to the use of such devices. This has caused some to dismiss the folks involved as a bunch of self-interested bomb makers trying to stir up business now that the Soviet threat has gone, but this is unfair. Atomic explosives are certainly humankind's most potent current physical capability. It's certainly reasonable to examine their practicality as a means of asteroid defense. Let's do that now.

Consider a one-kilometer-diameter asteroid heading toward the Earth at a typical interception speed of sixteen kilometers per second. There are many different types of asteroids. Some are made of iron-nickel and are as hard as steel. Others are made of stone. Others are made of weaker carbonaceous materials, and some even have a significant component of water ice. Let's say ours is a stone, since that is the most common type among the near-Earth group, and is midway in density between the iron and carbonaceous types. If this is the case, our asteroid will have a mass of about 2.5 trillion kilograms (2.5 billion tons). Now let's say we launch a ten-megaton H-bomb and detonate it right beside the asteroid so as to give it a sideways nudge. The bomb has a mass of about ten tons and releases 4×10^{16} joules (eleven terawatt-hours) of energy. If all of this energy goes into the bomb (i.e., none is lost by radiation), the fragments will explode with an average velocity of 2.8 million meters per second. The total impulse generated by the bomb will be 28 billion kilogram-meters per second, and one-quarter of this will be available to push the asteroid sideways. The asteroid will thus be imparted a velocity change of (7×10^9 kg-m/s ÷ 2.5×10^{12} kg) = 0.0028 m/s. Now the Earth has a diameter of approximately 12,800 kilometers, so we need to deflect the asteroid by about this much to make it miss. Dividing this distance by the 0.0028 m/s velocity increment (a first-order approximation to a calculation of trajectory change that is good enough for our rough estimation purposes here), we find that the bomb would need to be detonated 4.6 billion seconds, or 145 *years*, prior to impact to achieve the desired result. Chances are that we would not have that much time.

A much larger velocity change could be achieved if we inserted the bomb in a ground-penetrating warhead and fired it into the asteroid at high velocity, detonating beneath the surface. This would not work if the asteroid were a solid iron-nickel object, and it could fail even if the asteroid were mostly stone but had iron lumps in it that could destroy the bomb on impact. But if the asteroid were of the weaker stony or carbonaceous sort, good penetration should be possible. In this case, the 4×10^{16} joules represented by the bomb would be distributed not to ten tons of matter but to a much larger amount, perhaps a thousand tons. In that case, the characteristic velocity of the ejected mass would be reduced by a factor of ten compared to the previous calculation, but since there would be a hundred times as much mass ejected, the net result would be an impulse ten times as great. So now only a 14.5-year lead time would be needed. If the bomb's energy could somehow be usefully distributed across a hundred thousand tons of asteroid, the required lead time would fall to 1.45 years.

This sounds like it might be feasible, but wait. Where is the asteroid 1.45 years prior to impact? It is somewhere in deep space, with 730 million kilometers of travel path between it and the Earth. And how do we know how much mass the bomb will eject? To know that, we would have to know how deep it will penetrate after impact, which means we must know the geology of the object, and not just at its surface but underground as well. And how do we know the direction that fragmentation and ejection will occur? Subsurface strength variations could have a strong effect on the ejection vector. Once again, we need to know the subsurface geology in detail. And furthermore, the bomb will not only eject rocks, it will heat up the asteroid as well, possibly causing outgassing of volatile materials. The asteroid is probably rotating, so this outgassing will act to propel the asteroid in unanticipated directions unless the geometry, geology, and kinetic characteristics of the object are thoroughly understood.

There are too many unknowns. The fate of humanity is at stake in the success of the operation. If bombs are to be used as asteroid deflectors, they cannot just be launched willy-nilly. No, before any bombs are detonated, the asteroid will have to be thoroughly explored, its geology assessed, and subsurface bomb placements carefully determined and precisely located based on such knowledge. A human crew consisting of surveyors, geologists, miners, drillers, and demolition experts will be needed on the scene to do the job right.

But if a human crew is to be sent, there may be other ways besides bombs to give the asteroid the required push. For example, a spacefaring civilization will almost certainly develop space nuclear reactors with power outputs in the range of ten megawatts of electricity for the purpose of driving nuclear electric propulsion (NEP) ion-drive cargo vessels. Let's say we delivered one of these units to an asteroid and set it up it to drive a catapult firing chunks of the asteroid off into space at a velocity of 1 km/s. The catapult would thus act as a kind of rocket engine, using the asteroid's own mass as propellant. The average mass flow of the catapult would thus be about 20 kg/s, and the total thrust generated would be 20,000 newtons (equivalent to 4,490 pounds of force). This thrust would then be able to accelerate the asteroid in a precisely controlled direction at a rate of $(20 \times 10^3 \text{ newtons} \div 2.5 \times 10^{12} \text{ kg}) = 8 \times 10^{-9} \text{ m/s}^2$. This might seem imperceptible, but in the course of a year, a velocity increment of 0.25 m/s would be developed, sufficient to deflect the asteroid from colliding with the Earth, provided that the push was imparted at least 1.6 years in advance of the impact date.

Alternatively, if the asteroid has ice on it, this can be used as propellant in a nuclear thermal rocket (NTR). NTRs work by heating a working fluid to very high temperatures with a solid-core nuclear reactor and then ejecting it from a rocket nozzle as high-temperature gas. In the 1960s, the United States had a program called NERVA, which ground tested about a dozen NTRs with thrusts ranging from 45,000 newtons (10,000 pounds) to 1.1 million newtons (250,000 pounds) and power levels ranging from two hundred to five thousand megawatts. If hydrogen is used as the propellant, exploiting its low molecular weight to obtain high exhaust velocities, such engines operating at 2,500°C can generate specific impulse of 900 seconds (i.e., an exhaust velocity of 9 km/s). The 1960s NERVA engines actually generated about 825 seconds—almost twice that of the best chemical rocket engines possible. Wernher von Braun planned to use these engines for NASA's expeditions to Mars that were supposed to follow Apollo by 1981. Unfortunately, when the Nixon administration gutted the Apollo program and canceled plans for Mars, it derailed the NERVA program too, and the nuclear engines were never flown. However, the technology certainly works and, what's more, has an additional advantage beyond high performance: versatility. In principle, any fluid can be used as a propellant in an NTR. On Mars, carbon dioxide atmosphere is everywhere, so Mars-based NTR-powered rocket hoppers using CO_2 propellant could refuel themselves just by running a pump each time they land. Such "NIMF" vehicles, discussed in *The Case for Mars*, would give Martian explorers and settlers complete global mobility. Among the asteroids, ice is frequently available. This could be melted into water and stored in propellant tanks, then turned to steam thrust in the NTR engine. The specific impulse obtained would be about 350 seconds—nowhere near as good as hydrogen, but like their earthly counterparts, asteroid prospectors will need a mule that can live on mountain scrub, not a racehorse that eats only gourmet fodder. NTR steam rockets thus offer a very attractive technology for those wishing to get around among the asteroids. Furthermore, if an asteroid has enough ice, NTRs could be used very effectively to move it.

Let's say we took a five-thousand-megawatt NTR, no larger than the biggest NERVA engine tested in the 1960s, and placed it on a one-kilometer asteroid and fed it water propellant. The required mass flow would be 850 kilograms of water per second, and the thrust would be 2.9 million newtons (650,000 pounds). This would accelerate the asteroid at a rate of $(2.9 \times 10^6 \text{ newtons} \div 2.5 \times 10^{12} \text{ kg}) = 1.16 \times 10^{-6} \text{ m/s}^2$, or 36 m/s per year. This is more than a hundred times the acceleration possible with the electric catapult or bomb-driven systems discussed above. Using such technology, we could not only nudge the asteroid enough to make it miss the Earth on a particular pass, we could literally tug the asteroid into a substantially different orbit from which it would never threaten the Earth. We could even consider rearranging the orbits of asteroids so as to make groups of them more convenient for mining or other forms of development.

Furthermore, once fusion power is developed, fusion thermal rockets could be developed using deuterium from the asteroid's own water to provide the necessary energy, freeing such operations from need to obtain highly enriched uranium or plutonium from Earth.

So we need not be helpless in the face of asteroids. Two things are necessary for our defense: We must learn much more about the enemy. And we must become spacefarers.

ASTEROIDS AND LIFE

While averting doomsday asteroids is an important subject, ultimately it is merely a subset of a much more important endeavor, that of conquering the space frontier. Indeed, it is the absence of this insight that reduces much of the talk about the asteroid menace to mere alarmism.

Life on Earth has survived and prospered because at an early date it was able to take control, dictating the physical and chemical conditions of the planet in defiance of both solar and geological cycles. If it had not done this, terrestrial life would long since have gone extinct, as the sun today is more than 40 percent brighter than it was at the time of life's origin. Without life's ability to control terrestrial temperatures by regulating the CO_2 and other greenhouse gas content of the atmosphere, our ancestors would have all been cooked billions of years ago. Moreover, by replacing the CO_2 content of the atmosphere with oxygen, life transformed the terrestrial chemical environment to favor the development of species with the capability for increased activity and intelligence, and the ever more rapid evolution of still higher and more complex forms.

Within the history of the biosphere itself, the same phenomenon repeats. Those groups of species—whether natural ecosystems or human civilizations—whose activities effectively control their surrounding environments to favor their own growth are those that survive. Those which do not risk extinction. In the game of life, the only way to win is to have a part in making the rules.

On the short timescale, the relevant environment for most species is the Earth, and most ecosystems can get by for a fair while if they can deal with developments below the stratosphere. But over the long haul, this is not true. Since the success of the Alvarez hypothesis in explaining the mass extinctions that occurred at the end of the age of dinosaurs as having been caused by asteroid bombardment, it is now apparent that the relevant environment for life on Earth is not merely the planet of residence but the whole solar system.

Few people today understand this, yet it is true that subtle events in the asteroid belt determined the fates of their ancestors and may well in the future determine the fates of their descendants. It seems unbelievable that invisible happenings so far away could matter so much here. Similarly, throughout history, few inhabitants of rustic villages going about their daily lives were aware of the machinations of politicians and diplomats in their nations' capitals, which periodically would sweep the villagers off to die in cataclysmic wars on distant battlefields.

What you can't see can kill you. What you can't control probably will.

Humanity's home, humanity's environment, is not the Earth—it is the solar system. We've done well for ourselves so far by taking over the Earth and changing it in our own interests. Most people today, at least in the world's more advanced sectors, can walk about without fear of being dismembered by giant cats, are assured of sufficient food and fuel to survive the next winter, and can even drink water without risk of death. Sabertooths, locusts, and bacteria—for now, at least, we've beaten them all. But in a larger sense, we're still helpless. We may feel safe, having thrown our village's hoodlums in jail, but in the capital, behind the scenes, diplomats are meeting with generals, and plans are being made . . .

Our environment is the solar system, and we won't control our fate until we control it. The geological record is clear. Asteroids do hit. Mass extinctions, of sets of species every bit as dominant on Earth in their day as humanity is in ours, do occur.

You can't shoot down an incoming asteroid with an antiaircraft gun, air-to-air missile, or Star Wars defense system. If you want to prevent asteroid impacts, you have to be able to direct the course of these massive objects while they are still hundreds of millions of kilometers away from the Earth. A Type I civilization, however prosperous it might become, is intrinsically a helpless target in the asteroid shooting gallery. If it remains stunted at that level, its long-term prospects for survival are limited. If we are to be in charge of our fate, we must be able to control our *true* environment in the way that only a Type II civilization can: we must take charge of the asteroids.

In short, the lesson of the asteroids is this: If humanity wants to either progress *or* survive, we must become a spacefaring species. In the end, it is creativity, not austerity, that will be the key to our survival.

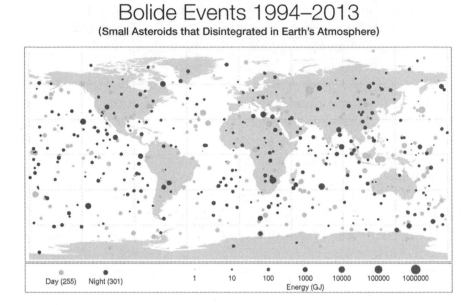

Bolide Events 1994–2013
(Small Asteroids that Disintegrated in Earth's Atmosphere)

Day (255)   Night (301)   1   10   100   1000   10000   100000   1000000
Energy (GJ)

Figure 11.3. Asteroid impacts are not ancient history. Earth continues to be bombarded by dangerous objects, with scores of smaller impacts over the past thirty years. (Note: 1,000 GJ = 200 tons of TNT.) *Image courtesy of NASA.*

Chapter 12

FOR OUR FREEDOM

The law of existence prescribes uninterrupted killing, so that the better may live.

—Adolf Hitler, 1941

Human civilization currently faces many serious dangers. The most immediate catastrophic threat, however, does not come from environmental degradation, resource depletion, or even asteroidal impact. It comes from *bad ideas*.

Ideas have consequences. Bad ideas can have really bad consequences.

The worst idea that has ever been is that the total amount of potential resources is fixed. It is a catastrophic idea, because it sets all against all.

Currently, such limited-resource views are quite fashionable among not only futurists but much of the body politic. But if they prevail, then human freedoms must be curtailed. Furthermore, world war and genocide would be inevitable, for if the belief persists that there is only so much to go around, then the haves and the want-to-haves are going to have to duke it out, the only question being when.

This is not an academic question. The twentieth century was one of unprecedented plenty. Yet it saw tens of millions of people slaughtered in the name of a struggle for existence that was entirely fictitious. The results of similar thinking in the twenty-first could be far worse.

The logic of the limited-resource concept leads down an ever more infernal path to the worst evils imaginable. Basically, it goes as follows:

1. Resources are limited.
2. Therefore, human aspirations must be crushed.

3. So, some authority must be empowered to do the crushing.
4. Since some people must be crushed, we should join with that authority to make sure that it is those we despise rather than us.
5. By getting rid of such inferior people, we can preserve scarce resources and advance human social evolution, thereby helping to make the world a better place.

The fact that this case for oppression, tyranny, war, and genocide is entirely false has made it no less devastating. Indeed, it has been responsible for most of the worst human-caused disasters of the past two hundred years. So let's take it apart.

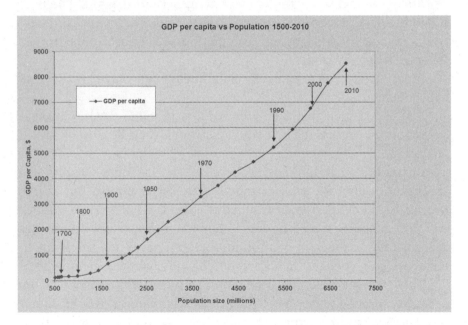

Figure 12.1. Contrary to Malthus's theory, human global well-being has increased with population size, and at an accelerating rate.

Two hundred years ago, the English economist Thomas Malthus set forth the proposition that population growth must always outrun production as a fundamental law of nature. This theory provided the basis for the cruel British response to famines in Ireland and India during the latter part of the nineteenth century, denying food aid or even regulatory, taxation, or rent relief to millions of starving people on the pseudoscientific grounds that their doom was inevitable.[1]

Yet the data show that the Malthusian theory is entirely counterfactual. In fact, over the two centuries since Malthus wrote, world population has risen sevenfold, while inflation-adjusted global gross domestic product per capita has increased by a factor of 50, and absolute total GDP by a factor of 350.

Indeed, it is clear that the Malthusian argument is fundamentally nonsense, because resources are a function of technology, and the more people there are and the higher their living standard, the more inventors, and thus inventions, there will be—and the faster the resource base will expand.

Our resources are growing, not shrinking, because resources are defined by human creativity. In fact, there is no such thing as "natural resources." There are only natural raw materials. It is human ingenuity that turns natural raw materials into resources.

Land was not a resource until people invented agriculture, and it is human ingenuity, manifested in continuous improvements in agricultural technology, that has multiplied the size of that resource many times over.

Petroleum was not originally a resource. It was always here, but it was nothing useful. It was just some stinky black stuff that sometimes oozed out of the ground and ruined good cropland or pasture. We turned it into a resource by inventing oil drilling and refining, and by showing how oil could be used to replace whale oil for indoor lighting, and then, later, by liberating humanity with unprecedented personal mobility.

This is the history of the human race. If you go into any real Old West antique store and look at the things owned by the pioneers, you will see things made of lumber, paper, leather, wool, cotton, linen, glass, carbon steel, maybe a little bit of copper and brass. With the arguable exception of lumber, all of those materials are artificial. They did not, and do not, exist in nature. The civilization of that time created them. But now go into a modern discount store, like Target. You will see some items made of the same materials, but much more made of plastic, synthetic fibers, stainless steel, fiberglass, aluminum, and silicon. And in the parking lot, of course, gasoline. Most of the materials that make up the physical products of our civilization today were unknown 150 years ago. Aluminum and silicon are the two most common elements in the Earth's crust. But the pioneers never saw them. To the people of that time, they were just dirt. It is human invention that turned them from dirt into vital resources.

There are things around today that clearly could become major resources but are not yet. Uranium and thorium were not resources at all until we invented nuclear power, but we are going to have to do a bit more inventing to get all the bugs out so as to unleash their truly vast potential. The same thing is true for solar energy, which needs to be made cheaper if it is to become truly practical as a baseload energy source. But this is happening, year by year, through innumerable inventions, great and small. Other enormous resources, more distantly in view, await the invention of ways to use them; for example, there is deuterium in seawater that could provide fusion power; there are methane hydrates and stratospheric winds. Today, the revolutionary new resource is shale. Twenty years ago, shale was not a resource. Today, as a result of the invention of new techniques of horizontal drilling and fracking, it's an enormous resource. In the past ten years, we've used it to increase US oil production 120 percent, from five million to eleven million barrels of oil per day. In the past twenty years, America's gas reserves have tripled, and we can and will do that and more for the world at large.

So the *fact* of the matter is that humanity is not running out of resources. We are exponentially expanding our resources. We can do this because the true source of all resources is not the earth, the ocean, or the sky. It is human creativity. It is people who are resourceful.

It is for this reason that, contrary to Malthus and all of his followers, the global standard of living has continuously gone up as the world's population has increased, not down. The more people—especially free and educated people—the more inventors, and inventions are cumulative.

Furthermore, the idea that nations are in a struggle for existence is completely wrong. Darwinian natural selection is a useful theory for understanding the evolution of organisms in nature, but it is totally false as an explanation of human social development. This is so because, unlike animals or plants, humans can inherit acquired characteristics— for example, new technologies—and do so not only from parents but from those to which they are entirely unrelated. Thus, inventions made anywhere ultimately benefit people everywhere. Human progress does not occur by the mechanism of militarily superior nations eliminating inferior nations. Rather, inventions made in one nation are transferred all over the world, where, newly combined with other technologies and different mind-sets, they blossom in radical new ways. Paper and printing were invented in China, but they needed to be combined with the Phoenician-derived Latin alphabet, German metal-casting technology, and European outlooks concerning freedom of conscience, speech, and inquiry to create a global culture of mass literacy. The same pattern of multiple sourcing of inventions holds true for virtually every important human technology today, from domesticated plants and animals to telescopes, rockets, and interplanetary travel.

Based on its inventiveness and its ability to bring together people and ideas from everywhere, America has become extremely rich, inciting envy elsewhere. But other countries would not be richer if America did not exist, or were less wealthy or less free. On the contrary, they would be immeasurably poorer.

Similarly, America would not benefit by keeping the rest of the world underdeveloped. We can take pride in our creativity, but in fact we would be much better off if all other people had as good a chance to develop and exercise their potential, and thus contribute to progress, as we do.

Nevertheless, so long as humanity is limited to one planet, the arguments of the Malthusians have the appearance of self-evident truth, and their triumph can have only the most catastrophic results.

Indeed, one has only to look at the history of the twentieth century, and the Malthusian/national social Darwinist rationale that provided the drive to war of both Imperial and, especially, Nazi Germany to see the horrendous consequences resulting from the widespread acceptance of such myths.

As the German General Staff's leading intellectual, General Friedrich von Bernhardi, put in his 1912 bestseller *Germany and the Next War*:

Strong, healthy, and flourishing nations increase in numbers. From a given moment they require a continual expansion of their frontiers, they require new territory for the accommodation of their surplus population. Since almost every part of the globe is inhabited, new territory must, as a rule, be obtained at the cost of its possessors— that is to say, by conquest, which thus becomes a law of necessity.[2]

Having accepted that war was inevitable, the only issue for the Kaiser's generals was when to start it, and they chose sooner rather than later so as not to give Russian industry a chance to develop.

Thus in 1914, the unprecedentedly prosperous European civilization was thrown into a completely unnecessary and nearly suicidal general war. A quarter century later, the same logic led the Nazis to do it again, with not merely conquest but systematic genocide as their insane goal.

To be perfectly clear on this point, the crimes of the Nazis were not just committed in secret by a few satanic leaders while the rest of the good citizens proceeded with their decent daily lives in well-meaning ignorance. In point of fact, such blissful ignorance was not possible. At its height, there were more than twenty thousand killing centers in the Third Reich, and most were discovered by Allied forces within hours of their entry into the vicinity—as the stench of their crematoria made them readily detectable. Something on the order of a million Germans were employed operating these facilities, and several million more were members of armed forces or police units engaged in or supporting genocidal operations.[3] Thus nearly every German had friends or family members who were eyewitnesses to or direct perpetrators of genocide, who could, and did, inform their acquaintances as to what was happening. (Many sent photos home to their parents, wives, or girlfriends, depicting themselves preparing to kill, killing, or posing astride the corpses of their victims.) Moreover, the Nazi leadership was in no way secretive about its intent; genocide directed against Jews and Slavs was the openly stated goal of the party *that eighteen million Germans voted for* in 1932. On March 20, 1933, less than two months after the Nazi assumption of power, SS leader Heinrich Himmler made it clear that these voters would have their wishes gratified, by announcing the establishment of the first formal concentration camp, Dachau, *at a press conference.* Furthermore, the implementation of the initial stages of the genocide occurred in public, with systematic degradation, beatings, lynchings, and mass murder of Jews done openly for all to see in the Reich's streets from 1933 onward, with the most extensive killings, such as those of the November 10, 1938, Kristallnacht pogrom, celebrated afterward at enormous public rallies and parties.

So the contention that the Nazi-organized Holocaust took place behind the backs of an unwilling German population is patently false. Rather, the genocidal Nazi program was carried out—and could only have been carried out—with the full knowledge and substantial general support of the German public. The question that has bedeviled the conscience of humanity ever since then: How could this have happened? How could the majority of citizens of an apparently civilized nation choose to behave in such a way? Some have offered German anti-Semitism as the answer. But this explanation fails in view of the fact that anti-Semitism had existed in Germany, and in many other countries such as France, Poland, and Russia to a sometimes much greater extent, for centuries prior to the Holocaust, with no remotely comparable outcome.

Furthermore, the Nazi genocidal program was not directed just against Jews but also at many other categories of despised people, including invalids, Romany people, and the entire Slavic race. Indeed, the Nazis had drawn up a plan, known as the Hunger Plan, for depopulating Eastern Europe, the Balkans, and the Soviet Union through mass starvation following their anticipated victory, on the insane supposition that by ridding the land of its farmers they could make more food available.[4] It should be noted that the partial implementation of this plan in occupied areas not only caused tens of millions of deaths but contributed materially to the defeat of the Third Reich, as it made it impossible for the Nazis to mobilize the human potential of the conquered lands on their own behalf. But not even such clear practical military and economic considerations could prevail against the power of a fixed idea.

In other words, as the Nazi leadership itself repeatedly emphasized, the genocide program was not motivated by mere old-fashioned bigotry. It certainly took advantage of such sentiments among rustics, hoodlums, and others to facilitate its operations. But it required something else to convince a nation largely composed of serious, solid, dutiful, highly literate, and fairly intellectual people to devote themselves to such a cause. It took Malthusian pseudoscience.[5]

Hitler himself was perfectly aware of the central importance of such an ideological foundation for his program of genocide. As noted Holocaust historian Timothy Snyder wrote in a September 2015 *New York Times* op-ed: "The pursuit of peace and plenty through science, he claimed in *Mein Kampf*, was a Jewish plot to distract Germans from the necessity of war."[6]

Once again, to be clear, the issue is not whether space resources will be made available to Earth in the proximate future. Rather it is how we, in the present, conceive the nature of our situation in the future. Nazi Germany had no need for expanded living space. Germany today is a much smaller country than the Third Reich, with a significantly higher population, yet Germans today live much better than they did when Hitler took power. So, in fact, the Nazi attempt to depopulate Eastern Europe was totally nuts, from not only a moral but also a practical standpoint. Yet, driven on by their false zero-sum beliefs, they tried anyway.

If it is allowed to prevail in the twenty-first century, zero-sum ideology will have even more horrific consequences. For example, there are those who point to the fact that Americans are 4 percent of the world's population yet use 25 percent of the world's oil. If you were a member of the Chinese leadership and you believed in the limited-resource view (as many do—witness their brutal one-child policy), what does this imply you should attempt to do to the United States?

On the other hand, there are those in the US national security establishment who cry with alarm at the rising economy and concomitant growing resource consumption of China. They project a future of "resource wars," requiring American military deployments to secure essential raw materials, notably oil.[7]

As a result of acceptance of such ideology, the United States has initiated or otherwise embroiled itself in conflicts in the Middle East costing tens of thousands of American lives (and hundreds of thousands of Middle Eastern lives) and trillions of dollars. For 1 percent of the cost of the Iraq War, we could have converted every car in the United States to flex fuel, able to run equally well on gasoline or methanol made from our copious natural gas.[8] For another 1 percent, we could have developed fusion power. Instead, we are fighting wars to try to control oil supplies that will always be sold to the highest bidder no matter who owns them.

Figure 12.2. Self-fulfilling prophecies. In 1912, the theoreticians of the German General Staff said it was inevitable that Germany would have to wage war for living space. In 2001, American geo-strategists proclaimed we would need to fight for oil. Both were wrong. Both set the stage for disaster. Much worse could be on the way if such zero-sum ideology is not discredited. *Image courtesy of Friedrich von Bernhardi*, Germany and the Next War, *trans. Allen H. Powles (London: Edward Arnold, 1918); Michael T. Klare*, Resource Wars *(London: Methuen, 1989); Graham Allison*, Destined for War *(Melbourne: Scribe Publications, 2018).*

There were no valid reasons for the first two World Wars, and there is no valid reason for a third. But there could well be one if zero-sum ideology prevails. Despite the bounty that human creativity is producing, there are those in America's national security establishment who today are planning for resource wars against peoples who could and should be our partners in abolishing scarcity. Their equivalents abroad are similarly sharpening their knives against us. This ideology threatens catastrophe.

Today there is a dangerous new anti-Western, antifreedom movement in Russia led by fascist philosopher Aleksandr Dugin, who is attempting to expand it worldwide. (He is doing so with significant success. The American "alt-right" and a host of similar European "identitarian" nativist movements all draw heavily from Dugin's ideas.[9] The basic idea is to both balkanize the West and undermine its commitment to humanist ideals by invoking the tribal instinct.) It is the contention of the Duginites that the world would be better off without America, or any other country with liberal values. Indeed, I was present at a conference on global issues held at Moscow State University, Dugin's home turf, in October 2013, when one of his acolytes got up and gave a fiery speech denouncing America for its profligate consumption of the world's resources, including its *oxygen* supply.[10] Such ideas amount to a call for war.

Do we really face the threat of general war? There seems to be no reason for it, and in fact, there isn't. People all over the world today are actually living much better than they ever did before, at any time in human history. But the same was true in 1914. Let us recall that a mere thirty years ago, the world was divided into two hostile camps, ready to spring into action on a few minutes' notice to destroy each other with tens of thousands of nuclear weapons. That threat vanished— not because of any change in real human circumstances, but due to the disappearance of a bad idea. It can just as quickly reappear with another. As in 1914 and 1939, all it takes is the *belief* that there isn't enough to go around—that others are using too much, or threatening by their growth to do so in the future—to set the world ablaze.

If it is accepted that the future will be one of resource wars, there are people of action who are prepared to act accordingly.

There is no scientific foundation supporting these motives for conflict. On the contrary, it is precisely because of the freedom and affluence of the United States that American citizens have been able to invent most of the technologies that have allowed China, Russia, and so many other countries to lift themselves out of poverty. And should China (with a population five times ours) develop to the point where its per capita rate of invention mirrors that of the United States—with 4 percent of the world's population producing 50 percent of the world's inventions— the entire human race would benefit enormously. Yet that is not how people see it, or are being led to see it by those who should know better.

Rather, people are being bombarded on all sides with propaganda, not only by those seeking trade wars, immigration bans, or preparations for resource wars, but by those who, portraying humanity as a horde of vermin endangering the natural order, wish to use Malthusian ideology as justification for suppressing freedom. Such arguments sometimes costume themselves as environmentalist, but that is deception. True environmentalism takes a humanist point of view, seeking practical solutions for real problems in order to enhance the environment for the benefit of human life in its broadest terms. It therefore welcomes technological progress. Antihuman Malthusianism, on the other hand, seeks to make use of instances of inadvertent human damage to nature as an ideological weapon of behalf of the age-old reactionary thesis that humans are nothing but pests whose aspirations need to be contained and suppressed by tyrannical overlords to preserve a divinely ordered stasis.

"The Earth has cancer and the cancer is man," proclaims the elite Club of Rome in one of its manifestos. This mode of thinking has clear implications. One does not provide liberty to vermin. One does not seek to advance the cause of a cancer.

The real lesson of the last century's genocides is this: We are not endangered by a lack of resources. We are endangered by those who *believe* there is a shortage of resources. We are not threatened by the existence of too many people. We are threatened by people who *think* there are too many people.

If the twenty-first century is to be one of peace, prosperity, hope, and freedom, a definitive and massively convincing refutation of these pernicious ideas is called for—one that will forever tear down the walls of the mental prison these ideas would create for humanity.

A QUESTION OF FAITH

We believe that free labor, that free thought, have enslaved the forces of nature, and made them work for man. We make the old attraction of gravitation work for us; we make the lightning do our errands; we make steam hammer and fashion what we need. . . . The wand of progress touches the auction-block, the slave-pen, the whipping-post, and we see homes and firesides and schoolhouses and books, and

where all was want and crime and cruelty and fear, we see
the faces of the free.
　　　　　　　　　　　—Colonel Robert G. Ingersoll,
　　　　　　　　　　　Indianapolis speech, 1876[11]

Western civilization is based on the radical individualist proposition advanced by the Greek philosophers Socrates and Plato that there is an innate faculty of the human mind capable of distinguishing right from wrong, justice from injustice, truth from untruth. Embraced by early Christianity, this idea became the basis of the concept of the *conscience*, which thereupon became the axiomatic foundation of Western morality. It is also the basis of our highest notions of law—the natural law determinable as justice by human conscience and reason, put forth, for example, in the US Declaration of Independence ("We hold these truths to be self-evident . . .")—from which we draw our belief that the fundamental rights of humans exist independent of any laws that may or may not be on the books or existing accepted customs. It is also the basis for *science*, our search for universal truth through the tools of reason.

As the great Renaissance scientist Johannes Kepler, the discoverer of the laws of planetary motion, put it, "Geometry is one and eternal, a reflection out of the mind of God. That mankind shares in it is one reason to call man the image of God." In other words, the human mind, because it is the image of God, is able to understand the laws of the universe. It was the forceful demonstration of this proposition by Kepler, Galileo, and others that let loose the scientific revolution in the West.

Science, reason, morality based on individual conscience, human rights; this is the Western humanist heritage. Whether expressed in Hellenistic, Christian, deist, or purely naturalistic forms, it all drives toward the assertion of the fundamental dignity of the human. As such, it rejects human sacrifice and is ultimately incompatible with slavery, tyranny, ignorance, superstition, perpetual misery, and all other forms of oppression and degradation. It asserts that humanity is capable and worthy of progress.

This last idea—progress—is the youngest and proudest child of Western humanism. Born in the Renaissance, it has been the central motivating idea of our society for the past four centuries. As a civilizational project to better the world for posterity, its results have been spectacular, advancing the human condition in the material, medical, legal, social, moral, and intellectual realms to an extent that has exceeded the wildest dreams of its early utopian champions.

Yet now it is under attack. It is being said that the whole episode has been nothing but an enormous mistake, that in liberating ourselves we have destroyed the Earth. As influential Malthusians Paul Ehrlich and John Holdren put it in their 1971 book *Global Ecology*:

> When a population of organisms grows in a finite environment, sooner or later it will encounter a resource limit. This phenomenon, described by ecologists as reaching the "carrying capacity" of the environment, applies to bacteria on a culture dish, to fruit flies in a jar of agar, and to buffalo on a prairie. It must also apply to man on this finite planet.

Note the last sentence: *It must also apply to man on this finite planet.* Case closed. The only thing left to decide is who gets death and who gets jail.

We need to refute this. The issue before the court is the fundamental nature of humankind. Are we destroyers or creators? Are we the enemies of life or the vanguard of life? Do we deserve to be free?

Ideas have consequences. Humanity today faces a choice between two very different sets of ideas, based on two very different visions of the future. On the one side stands the antihuman view, which, with complete disregard for its repeated prior refutations, continues to postulate a world of limited supplies, whose fixed constraints demand ever-tighter controls upon human aspirations. On the other side stand those who believe in the power of unfettered creativity to invent unbounded resources and so, rather than regret human freedom, demand it as our birthright. The contest between these two outlooks will determine our fate.

If the idea is accepted that the world's resources are fixed with only so much to go around, then each new life is unwelcome, each unregulated act or thought is a menace, every person is fundamentally the enemy of every other person, and each race or nation is the enemy of every other race or nation. The ultimate outcome of such a world-view can only be enforced stagnation, tyranny, war, and genocide. Only in a world of unlimited resources can all men be brothers.

On the other hand, if it is understood that unfettered creativity can open unbounded resources, then each new life is a gift, every race or nation is fundamentally the friend of every other race or nation, and the central purpose of government must not be to restrict human freedom but to defend and enhance it at all costs.

It is for this reason that we need urgently to open the space frontier. We must joyfully embrace the challenge of launching a new, dynamic, pioneering branch of human civilization on Mars—so that its optimistic, impossibility-defying spirit will continue to break barriers and point the way to the incredible plentitude of possibilities that urge us to write our daring, brilliant future among the vast reaches of the stars. We need to show for all to see in the most sensuous way possible what the great Italian Renaissance humanist Giordano Bruno boldly proclaimed: "There are no ends, limits, or walls that can bar us or ban us from the infinite multitude of things."

Bruno was burned at the stake by the Inquisition for his daring, but fortunately others stepped up to carry the banner of reason, freedom, and dignity forward to victory in his day. So we must do in ours.

And that is why we must take on the challenge of space. For in doing so, we make the most forceful statement possible that we are living not at the end of history but at the beginning of history; that we believe in freedom and not regimentation, in progress and not stasis, in love rather than hate, in peace rather than war, in life rather than death, and in hope rather than despair.

Chapter 13

FOR THE FUTURE

As I surveyed them from this point, all the other heavenly bodies appeared to be glorious and wonderful,—now the stars were such as we have never seen from this earth; and such was the magnitude of them all as we have never dreamed; and the least of them all was that planet, which farthest from the heavenly sphere was shining with borrowed light. But the spheres of the stars easily surpassed the earth in magnitude—already the earth itself appeared to me so small, that it grieved me to think of our empire, with which we cover but a point, as it were, of its surface.

And as I gazed upon this more intently, "Come!" said Africanus, "how long will your mind be chained to the earth? Do you see into what regions you have come?"
—Marcus Tullius Cicero, *The Dream of Scipio*, 51 BCE

What kind of future can we create?

Homo sapiens began its existence 200,000 years ago as a handful of tribes living in the Kenyan Rift Valley. There, in our natural habitat, we remained for the following 150,000 years, under conditions of near-complete technological stagnation. But then, for some reason, 50,000 years ago, some of our ancestors ventured forth, to take on the more challenging environments of Ice Age Europe and Asia, diversifying and inventing new ways of doing things as they went, to ultimately settle the entire Earth. The trek out of Africa was humanity's key step in setting itself on the path toward achieving the mature Type I status that the human race now approaches.

The challenge today is to move on to Type II. Indeed, the establishment of a true spacefaring civilization represents a change in human status fully as profound—both as formidable and as pregnant with promise—as humanity's move from the rift valley to its current global society.

Space today seems as inhospitable and as worthless a domain as the wintry wastes of the north would have appeared to a denizen of East Africa fifty thousand years ago. Yet, like the north, it is the frontier arena whose possibilities and challenges will allow and drive human society to make its next great positive transformation. If we take it on, we can create a future as grand and magnificent in its prospects compared to our current global society as it is in comparison to that of our ancestral tribes huddled in their caves in a small part of East Africa. From the standpoint of that future, nothing else being done today is of remotely comparable importance. Our time will be remembered because this is when humans first set sail for other worlds.

The task for our time is to open the way to Type II. But we should understand where the road leads. We are bound for the stars.

Columbus dared the Atlantic in small, frail coastal craft that, even fifty years later, no one would have attempted to use to cross the ocean. This was how it had to be. Until European civilization became transatlantic, there was no cause to develop truly Atlantic-capable vessels. But once Europeans found their New World, soon enough they developed reliable three-masted caravels, then clipper ships, steamships, ocean liners, and Boeing 747s.

Similarly, the first explorers will make their six-month voyages to Mars in small, cramped vessels, demonstrating a toughness that will be a source of awe to their grandchildren, who will do their three-week interplanetary transits in style aboard well-appointed fusion-powered spaceliners. But the same technology that makes travel to Mars a matter of ease and comfort for everyone will make it possible for those with the bravest spirits to venture much further.

A century ago, the Russian space visionary Konstantin Tsiolkovsky famously said, "The Earth is the cradle of mankind, but one cannot live in the cradle forever."[1] Indeed. The Earth has been our cradle, within which we have developed the capability to enter the solar system, which shall be our schoolyard. There we shall grow bigger, stronger, smarter, wiser, and braver, preparing us to enter the universe at large.

So we will continue to move out.

THE YEAR 2069

As I write these lines, the fiftieth anniversary of the Apollo moon landing is coming into view. Over the past fifty years, our robotic planetary program has performed epic deeds of exploration, while our human spaceflight effort has stagnated. But now, with the entrepreneurial space launch revolution, we are poised to break out into the solar system. If we seize this opportunity, where might we be fifty years hence?

Here is my vision of where we could be. We will have fusion power and open-sea mariculture and will no longer be living in fear of climate change, resource exhaustion, or each other. We will be a cosmopolitan civilization, able to travel the globe freely through suborbital space in less than an hour, so that nearly everyone will have friends in every land. We will have research laboratories, industries, and hotels on orbit. We will have scientific bases, astronomical interferometers, and helium-3 mines on the moon. There will be operational lunar skyhooks, enabling transport all over the moon, and the cheap lifting of propellant to lunar orbit to support exploration missions to the outer solar system. We will have city-states on Mars—vibrant optimistic centers of invention, sporting lively and novel cultures, with many casting off the chains of tradition to strike out new paths to show the way to a better future. We will have mining and settlement outfits finding their way into the main asteroid belt, and exploration expeditions to the outer solar system to test the means by which we might access its enormous energy resources for the human future. We will have grand observatories floating in free space that will be making magnificent discoveries in physics and cosmology, mapping the planets of millions of stars, and finding other worlds filled with life and intelligence. We will be learning the truth about the nature of the universe, and life's role in it, and preparing our first interstellar spaceships to journey forth and find our place among the stars.

This is what we can do, we who are alive today. This is the future *we* can make. This is the grand heritage that we can pass on to those who will go further.

And then?

TWELVE THOUSAND BY 3000

Ultimate limits are impossible to even conceive, let alone predict. So let's adopt as our time horizon for the future a date a thousand years hence. What might human civilization be like in the year 3000?

Virtually any projection made across such a span of time will no doubt prove to be radically conservative. No one from two centuries ago, let alone ten, could possibly anticipate—or even believe if he or she were told—the everyday features of our civilization today, including as it does thousands of giant ships hurtling through the air at 500 miles per hour and instantaneous communication around the globe. Even a few centuries from now, our present ideas about future spaceflight will probably appear far more quaint than those of the best nineteenth-century visionaries do to us today. Writing in 1865, Jules Verne told how humans could go to the moon.[2] He predicted that it would be Americans who would do it; that they would launch from Florida; and that the crew would consist of three men, who would travel in a capsule, orbit the moon, land in the Pacific Ocean, and be picked up by a US Navy warship, all as actually happened 104 years later. But the propulsion system he employed was heavy artillery. This was perhaps a natural mistake for him to make, because big guns were the most potent instruments of his time. But it was wrong, not only because of technicalities concerning the performance limits of artillery propellants and acceptable g-loads for crew, but for the more fundamental reason that he was a nineteenth-century mind grappling with a twentieth-century problem. Similarly, we today can talk about interstellar settlement using fusion-powered spacecraft delivering colonists who terraform planets by building fluorocarbon factories, because such technologies are within our engineering horizon. But we are twenty-first-century minds (I'm actually a twentieth-century mind) dealing with a twenty-third-century problem. If this book survives the ages, readers in the Tau Ceti system several centuries from now may remark how insightful it was for someone living in such a primitive time as ours to foresee interstellar colonization. "But doing it with thermonuclear interstellar spaceships and greenhouse gas factories," they might say, "how twentieth-century can you get? Of course, they couldn't imagine that we would do it using laser-projected self-replicating nanomachines and programmable microbes. . . ."

While we can imagine such possibilities, we can imagine magic too. If we are to stay within the world of speculative engineering, as opposed to science fiction or fantasy, we need to stay within science as it is known today, even if that necessarily means that we will wildly underpredict the possibilities.

Well then, using fusion-powered spacecraft, spacecraft speeds of 5–10 percent the speed of light seem achievable. If we embrace the Noah's Ark Egg concept, or something like it, then 20 percent light speed could be in reach. In our region of the galaxy, stars are typically spaced about six light-years apart, with a density of about one every three hundred cubic light-years. So allowing a century for a twenty-light-year expansion step and another century to develop the new frontier planet to the point where it becomes the launching point for the next move, it would appear that the velocity of our settlement wave could potentially reach something on the order of 10 percent the speed of light. This being so, then a thousand years from now, human civilization could encompass a sphere a hundred light-years in radius surrounding the Earth. Within this zone, there are twelve thousand stars.

What might we do with twelve thousand stars?

In our neighborhood of the galaxy, about 3 percent of them are likely to be bright yellow-white type F stars, 7.5 percent sunlike type G stars, 12 percent type K orange dwarfs, and 76 percent smaller type M dwarfs. All will have habitable zones, near or far, depending on the brightness of the star. Nearly all are likely to have at least several planets, accompanied by scores of moons and innumerable asteroids.

The majority of the planets or moons found in the habitable zones probably won't be ideal places for settlement when first encountered. But a civilization capable of interstellar flight will be more than capable of terraforming. Based on our experience terraforming Mars, Venus, the asteroids, and the moons of the outer planets, we will move out and make dead worlds come alive wherever we go. We will seed thousands of new created worlds with fabulous arrays of new species. We will advance and multiplex evolution. We will turn geospheres into biospheres, and biospheres into noospheres.[3] We will found twelve thousand new branches of human civilization.

What will they be like? Unquestionably, they will be very techno-logically advanced. Inventions are cumulative, and inventions made anywhere ultimately benefit everyone everywhere. Assuming an average population of ten billion people per solar system, with typical living standards and levels of education greatly exceeding anything seen today, humanity's inventive force will number in the tens of tril-lions and produce staggering results. With thousands of unique civili-zations contributing and exchanging their discoveries via interstellar radio transmission, an immense storehouse of knowledge will be created, advancing and expanding at a stupendous rate. We will plumb the secrets of nature. We will find the cures to all diseases and infirmi-ties. We will improve ourselves. We will become different, in innumer-able different ways.

This is a very good thing, because it is essential that humanity become more diverse.

In the twenty-first century, with the advent of global communica-tions, jet aircraft, and, soon, rocket planes, the potential for terrestrial geographic barriers to maintain cultural diversity is in the process of being eradicated. In consequence, the world is now rapidly being homogenized to a single culture, and this tendency will only accel-erate. Were we to remain earthbound, we would soon not be only one species but one culture. If studies of biology and evolution are any guide, that is a prescription for disaster.

However, fortunately, the same general level of technology that makes the Earth too small to maintain human diversity is now opening up a much wider sphere for our development. This should generate conditions for the rapid regeneration of both cultural and biological diversity, across a vastly broader theater of possibilities. In nature, large interconnected gene pools are very slow to evolve, as it takes a very long time for any new trait to become dominant. In contrast, the generation of new species is favored when a small group becomes isolated from the main stock of its kind and put in a new environment where it is subject to novel, adaptive stresses. Under such conditions, the generation of new traits is called forth, and since the genes for the new trait are not being constantly washed out by interbreeding with the primary population, new varieties and new species rapidly result. Analogous processes of innovation-in-isolation are necessary for the generation of substantially new cultures by human populations.

As humans expand to Mars, the asteroids, the outer solar system, and ultimately the stars, precisely such conditions for first cultural and, ultimately, genetic diversification will be obtained. Culture will change first, then language.

Among the first things that will change will be forms of political organization. Despite all the progress that has been made over centuries of struggle, it is apparent that the well of human thought on the best forms of social organization has not been exhausted by the present age. We still have a long way to go before we create just societies that truly maximize the human potential of every individual. Even the most advanced Western nations still have a class system, using requirements for expensive university degrees or other certifications as poll taxes to deny access to professions to those who would be kept down. We keep millions in prisons, artificial hells that we, with fabulous dishonesty, label "correctional institutions." Our educational system has created universal literacy, but at a low level, and is hostile to creativity. Popular culture is a degraded mess. We spend far too much on war and too little on science. Our legal system protects the strong but does poorly for the weak. Political power is divided between a corrupt political class, an arrogant bureaucracy, and a small group of the superrich. We can certainly do better.

There is no consensus on how these problems can be solved. But as humanity moves out into the solar system, those with new ideas on how we might do better will have a chance to give their diverse beliefs a try. This opportunity to be the maker of one's world, instead of a mere inhabitant of one already made, is a fundamental form of human freedom, and people are sure to seize it once it becomes available. So, Type II will see the establishment of many new branches of human civilization on the moon, Mars, the asteroids, the satellites of the outer planets, and even the Kuiper Belt, organized in accordance with myriads of divergent social and political concepts. Most of the truly novel ones may fail because, for the most part, things are done now the way they are done for a reason. Tyrannies will fail, as they always do. But some of the new worlds will find a better way to maximize human potential. These will draw immigrants, prosper, grow, expand, and then be emulated by others, as a result. In this way, interplanetary diversity will lead to social progress.

After a period of such diverse experimental development, a Type II civilization might ultimately become politically unified, but a Type III civilization cannot. The distances between the stars are simply too great for any sort of enforcement. So there will never be an interstellar empire. Rather, there will be a vast and expanding collection of Type II city-states, diverging in culture, social, and political organization, but trading thoughts, and learning from the successes or failures of each.

While all the major languages on Earth have a recognizable history going back millennia, they have changed a lot. Many people today have trouble with the English of Shakespeare, and few can understand that of Chaucer—who wrote a mere seven hundred years ago—without university training. The English may owe their existence as a nation to Alfred the Great, but they can no longer understand him. So there will be new languages—thousands of them—new literatures, and new poetry.

But things will likely go much further. Due to the intrinsic enormous differences in environment from one extraterrestrial habitat to another (down to things as fundamental as the gravitational field within which a civilization exists), it is certain that not only culture but heredity will be driven fast and hard in many diverse directions.

It might be maintained that in the future, the increasing human ability to control heredity will impede this process. I would argue the contrary. In fact, since cultural evolution occurs normally on a much faster timescale than genetic evolution, as soon as human beings gain control of the genetic code—that is, culture gains control of heredity—biological evolution will occur at a greatly accelerated pace. It might be the case that in one locality or another, governments will act to suppress this kind of self-directed evolution. However, enforcement of any sort of government edicts across interstellar space is likely to be impossible. Therefore, among the culturally diverse civilizations, some might choose to suppress change, but others will drive it. The difference of opinion on this score will thus only serve to accelerate the process of multiple speciation.

One of the results of these programs of self-modification will no doubt be a drastic extension of the individual human life span. The aging process itself may well be defeated. If that is the case, however, the necessity for space expansion will be greatly accentuated, as the younger generation will face old worlds in which the determining roles have already been assigned. Humans need to matter. In the age of quasi-immortality, new generations will need new worlds to give their lives immortal purpose. Fortunately, long-lived people will be able to undertake long voyages. Thus the interstellar diaspora, and its production of ever more diversity, will be driven even further.

In science fiction television and film, such as *Star Trek*, the galaxy is frequently depicted as being inhabited by numerous species of "aliens" who are humanoid in all respects except for minor differences, such as skin color or ear shape, and who can sometimes interbreed. Humans are the product of four billion years of terrestrial evolution, and while for reasons of convergent evolution, it is possible that aliens might be found with a general form similar to humans (i.e., two arms/two legs/ head on top is a fairly practical design plan), they would obviously differ enormously from us internally at every level, including organs, tissues, and cellular structure. Interbreeding would thus be clearly an impossibility. However, if the process of human diversification alluded to in the previous paragraph were to go forward, then it is highly probable that a thousand years from now, interstellar space in this region of the galaxy will be populated by numerous human-descended intelligent species that will differ from each other in appearance, emotional makeup, and other characteristics to a considerably greater extent than the cosmopolitan species that populate the *Star Trek* universe. Local stylistic fads could well create races with green skin or pointed ears; more serious considerations such as gravity differences might drive the development of outlandishly tall-thin (low gravity) or short-powerful (high gravity) varieties. Many of them, as a result of their self-directed evolution, will be far more intelligent, sensitive, healthy, long-lived, athletic, graceful, and (to themselves, anyway) beautiful than we.

But, in their own way, they will all be human.

Eventually, however, we will meet people who aren't. This is so because, as we discussed earlier, the fraction of biospheres sporting civilizations is likely to be given by the ratio of the lifetime of a species-civilization to the time it takes a mature biosphere to generate an intelligent species. We have limited data to judge this, but going on the basis of the experience of our own planet, this is likely to be on the order of one in a thousand. So, as humanity expands to tens of thousands of solar systems, the probability of direct contact with alien civilizations will increase to near certainty.

What then? Might they not be more advanced than us?

They certainly might. But I think we'll be fine regardless. While alien invasions are a staple of science fiction, the logistics of interstellar warfare provide enormous advantages to the defense, as the home team is likely to outnumber the visitors by millions to one. Provided they have attained a minimally competent Type II status, the defenders would almost certainly be able to vastly outgun any invading force as well.

But there will be no war between starfarers, because no species will ever be able to make it to Type III without understanding the universal beneficial role of creative intelligence.

It is intelligence that creates resources. The more creators, the more resources.

If they are wise enough to avoid destroying themselves while possessing Type II technologies, they'll be wise enough to know that. The same will be true of us.

So I believe we will meet them as friends, and their friends as well, with great benefit to all as the circle expands, as each will be able to acquire from the rest not only vast knowledge but entirely new ways of understanding.

Perhaps there already is such a Galactic Club. If so, we need to prepare ourselves to join it. If not, we need to prepare ourselves to start it.

Type III calls.

CONCLUSION

Our distant kin followed the stars to the north. Later, as humans became seafaring, it was the stars again—with poetic truth, the North Star—that gave us the guidance we needed to become a truly global species.

Today the stars beckon again, this time not to new continents but to new solar systems. Multitudes of new worlds yet unknown await, filled with menaces to be faced, challenges to be overcome, wonders to be discovered, and history to be made. The first chapter of the human saga has been written, but vast volumes lying out among the stars are still blank—ready for the pens of new peoples with new thoughts, new tongues, astonishing creations, and epic deeds.

It is a grand time to be alive. We are young, the universe is in its spring, and the door has been opened, inviting us outside to meet the dawn of the greatest adventure ever.

Ad Astra.

Chapter 14

WHAT NEEDS TO BE DONE

We are living at a moment of great promise. Humanity's breakout into space is underway. Yet success is not inevitable. There is no such thing as destiny. Things happen because people make them happen. If what needs to be done is left undone, magnificent futures can die stillborn in the womb of time. Commenting on the failure of the French Revolution to produce a free society, the German writer Friedrich Schiller said, "A great moment found a little people." We don't want a future historian to someday say the same thing about us.

It is not enough to cheer the efforts of those currently in the arena. Musk, Bezos, and the rest could easily fail. They need reinforcements. If you have the technical or business skills to help them, you might consider enlisting under one of their banners—or, better yet, think about joining with others to start new space ventures. It's not just space launch that needs revolutionary technology. There are plenty of problems that have to be solved to fully open the space frontier, and plenty of opportunity—and need—for new players with new approaches and new ideas.

A century and a half ago, Horace Greeley advised Americans, "Go West, young man, Go West." He was right. If you want to do something grand with your life, the frontier is the place to be. Today, the frontier is not the West but the sky. So, look up, young minds, look up.

But whether or not you are personally qualified or situated to join the pioneers, there is still plenty you can do. Many things are progressing well in space right now, but many are not, and there are important political and cultural issues that remain to be resolved, which must be resolved favorably if the spaceflight revolution is to fully bear fruit.

A PROGRAM FOR ACTION

In the beginning, there was the word.

There are those who think that because the entrepreneurial space companies like SpaceX and Blue Origin are moving ahead so nicely, we no longer need NASA or other government-led efforts. They could not be more mistaken. There are commercial opportunities that can support private space activities in suborbital and geocentric space, but they will need public support to make sure they are not blocked by hostile or obtuse bureaucracy. Moreover, the critical initial breakout to the moon, Mars, and beyond will need government funding. This is consistent with the history of exploration and settlement on Earth, where high-risk first missions like those of Columbus and Lewis and Clark needed government backing, with commercial development following later. The space entrepreneurs are facilitating the launch of such initiatives, by developing a substantial fraction of the required flight hardware set in advance. This is dramatically lowering the cost, risk, and schedule thresholds associated with such programs, thereby making them much more attractive to the political class, and more sustainable as well. But still, such a decision will need to be obtained.

It's going to take a public-private partnership to place humanity on the moon and Mars. Right now, the private side of that partnership is advancing boldly. But the equally necessary public side—the space program that reports to you and me—is badly adrift.

NASA deserves a lot of credit. A space agency funded by 4 percent of the world's population, it is responsible for launching 100 percent of all the rovers that have ever wheeled on Mars; all the probes that have visited Jupiter, Saturn, Uranus, Neptune, or Pluto; nearly all the major space telescopes; and all the people who have ever walked on the moon. But while its robotic planetary exploration and space astronomy programs continue to produce epic results, for nearly half a century its human spaceflight effort has been stuck in low Earth orbit. The reason for this is simple: NASA's space science programs accomplish a lot because they are *mission driven*. In contrast, the human spaceflight program has allowed itself to become *constituency driven* (or, to put it less charitably, *vendor driven*). In consequence, the space science programs spend money in order to do things, while the human spaceflight program does things in order to spend money. Thus, the efforts of the science programs are focused and directed, while those of the human spaceflight program are purposeless and entropic.[1]

This was not always so. During the Apollo period, NASA's human spaceflight program was strongly mission driven. We did not go to the moon because there were three random constituency-backed programs to develop Saturn V boosters, command modules, and lunar excursion vehicles, which luckily happened to fit together, and which needed something to do to justify their funding. Rather, we had a clear goal—sending humans to the moon within a decade—from which we derived a mission plan, which then dictated vehicle designs, which in turn defined necessary technology developments. That's why the elements of the flight hardware set all fit together. But in the period since, with no clear mission, things have worked the other way.

Neither the space shuttle nor the International Space Station were designed as parts of any well-conceived plan to send humans to the moon or Mars. So, like a ballerina demanding a rewrite of *Macbeth* to display her skill, insistence that they be included as part of such programs only served to make them infeasible. More recently, other constituencies in NASA have made demands that any expedition to the moon or Mars make use of new hobbyhorses, including variously a space station or asteroid fragment in lunar orbit or high-powered electric propulsion, none of which are necessary, desirable, or, arguably, even acceptable, for near-term human exploration.

The current Deep Space Gateway (aka "Lunar Orbit Platform-Gateway," or "LOP-G"—I am not making this up) program is a case in point.[2] If you want to understand the merit of this project, consider a business proposition where you are offered a chance to rent an office in Saskatoon. Under the terms proposed, you will need to pay to build the office building and agree to a thirty-year lease at $100,000 per month rent, with no exit clause. In addition, you will need to spend one month per year in Saskatoon and travel through Saskatoon on your way to anywhere else for the rest of your life. That, in a nutshell, is the DSG/LOP-G project. It will cost a fortune to build and a fortune to maintain, and it will add to the propulsion requirements and timing constraints of all missions to the moon and Mars that are forced to stop there—as they surely will, since otherwise the pointlessness of building it will be revealed to the public. It is not an asset but a liability, or rather an entitlement, created for no other purpose but to provide a mechanism to drain agency funds to NASA's largest contractors.

This is unacceptable. NASA's space program is *our* space program. It does not belong to the major aerospace contractors, or even to NASA's management. It belongs to *us*. That some of the money that NASA's human spaceflight office throws around on useless projects might end up in the hands of entrepreneurial space companies is not enough. The American people deserve a space program that is really going somewhere. We are paying for it. We have a right to insist on real results.

The mission needs to come first. First and foremost, the NASA human spaceflight program needs a clear, driving goal, which should be to initiate a permanent human presence on the moon and Mars within a decade. Such a deadline is as necessary as a defined destination, because without it, the goal has no force and activities will continue to be directed by entropic vendor or political constituency pressure, rather than by the alleged purpose.

Rather than continue paying for endless cost-plus contracts to "develop" things with no real purpose, NASA needs to set clear goals and contract for services to support those goals. So, for example, let's say enabling human lunar exploration is the goal—as it currently allegedly is. NASA should put out a request for proposals to industry for systems to deliver cargos to the moon, and astronauts round-trip, offering to match development costs dollar for dollar and to award a certain number of missions to the best bidders. Whoever got such a contract would be strongly incentivized to minimize development cost and time because they would be paying half the cost out of pocket and would not start making a profit until actual missions began. This, in fact, is how the Commercial Orbital Transportation Services (COTS) program set up by former NASA administrator Mike Griffin enabled the rapid development of the Dragon system for delivering first cargo, and now crew, to the International Space Station at a cost to the agency of less than 5 percent of what it has thus far spent on the cost-plus

Orion system—which, after fifteen years of development, has yet to be flown.

If NASA wants to send humans to the moon or Mars, it should not spend billions on random cost-plus infrastructure projects that supposedly might come in handy if someday there were a program to go. Instead it should just take competitive bids for delivery services. It should incentivize the development of additional systems, including rovers, habitats, life support, power units, space suits, and so on, the same way.

Approached in this way, we can have our first permanent bases established and operating on both the moon and Mars within a decade, for a small fraction of NASA's current budget. We will also have a vibrant private space industry, driving down the cost and advancing the technology of launch vehicles, spacecraft, propulsion, and every other system needed for space exploration and development with all the ferocious creativity that free enterprise can bring to bear. With that, the doorway to the universe will be flung wide open.

Europe, Russia, China, Japan, India, and others could do the same thing, provided they choose to be part of humanity's future. I hope they do. There is room in space for more than one flag.

But a decision to do it is required. That is where you come in.

WHAT YOU CAN DO

If you want humans to get to Mars, then you need to become a space activist.

We have immense latent support for space exploration in America and many other countries, but only a tiny fraction of it is organized. The harvest is plentiful, but the gatherers are few. Permanent organizations with active memberships are needed to generate the kind of political muscle required. In a nutshell, Mars needs you. It's not enough to wish the space program well; if you believe in a future that is not limited by Earth's horizons, you need to join with other like-minded individuals and make your voice heard. Joining a space activist organization is probably the best way to do that.

In the United States, there are basically four organizations to choose from. I'm a bit prejudiced here because I happen to be a leader of one of them, the Mars Society. But I'll try to give you an accurate enough picture to decide where you should center your efforts.

The Planetary Society is the largest of the four, with perhaps fifty thousand members. Founded by Carl Sagan, Louis Friedman, and former Jet Propulsion Lab director Bruce Murray, it is led today by astronomer Jim Bell and television science educator Bill Nye. The Planetary Society is primarily interested in promoting the robotic exploration of the solar system, but it is supportive of a humans-to-Mars program, provided it is done along the lines of Carl Sagan's international collaboration model. You can join the Planetary Society by sending a check for $37 to The Planetary Society, 85 South Grand, Pasadena, CA 91105. For more information, go to www.planetary.org.

The National Space Society is the second largest, with twenty thousand members. Founded by Wernher von Braun and Princeton space visionary Professor Gerard O'Neill, it is led today by Mark Hopkins, Kirby Ikin, and Alice Hoffman. The primary interest of the NSS is to promote the human settlement of space, including the moon, Mars, the asteroids, and free-floating space colonies. The NSS would be equally happy supporting a moon or Mars program based on any of the patriotic-JFK, internationalist-Sagan, or private-enterprise-led models. The NSS is organized into about a hundred local chapters, which organize local and regional events as well as a national conference once a year. You can join the NSS by sending a check for $20 to National Space Society, 1155 15th Street NW, Suite 500, Washington, DC 20005. Membership benefits include a bimonthly glossy magazine and frequent mobilization bulletins concerning the space program. For more information, go to www.nss.org.

The Space Frontier Foundation is the smallest organization, with about five hundred members. Founded by Rick Tumlinson and Jim Muncy, and led today by Jim Feige and Charles Miller, the Space Frontier Foundation has a very strong free-enterprise tilt. Of the three approaches to space mentioned above, it would favor only the free enterprise model. If opening space with maximum free enterprise and minimum government involvement is fundamental to your principles, then consider joining this group. The Space Frontier Foundation sponsors one national conference per year. You can join the Space Frontier Foundation by sending $25 to The Space Frontier Foundation, 16 First Avenue, Nyack, NY 10960. For more information, go to www.spacefrontier.org.

The Mars Society is the newest of the space organizations. Together with many other members of the Mars Underground, including Chris McKay, Carol Stoker, and Tom Meyer, as well as science fiction authors Greg Benford and Kim Stanley Robinson, I founded the Mars Society with the purpose of furthering the exploration and settlement of Mars by both public and private means. Our founding convention in Boulder, Colorado, in August 1998 drew seven hundred people from forty countries, featured 180 papers and talks on everything from Mars mission strategies to the ethics of terraforming, and attracted international press coverage. As of this writing, we have around seven thousand members, divided into seventy chapters, of which forty are in the United States and thirty span the globe. Our activities include broad public outreach, political lobbying, and the operation of two human Mars exploration simulation bases: one in the polar desert on Canada's Devon Island and the other in the desert of southern Utah. To date, more than two hundred crews of six people each have gone to these stations and undertaken simulated Mars missions ranging from two weeks to four months in duration. During these missions, crews are tasked to perform sustained programs of field exploration in geology and microbiology while operating under many of the same constraints that human explorers would face on Mars. By doing so, we are learning a great deal about what field tactics and technologies will prove most useful when humans finally journey to the Red Planet. At the same time, press reportage of these missions—which has appeared in the world's leading media, ranging from the *New York Times*, CNN, and the Discovery Channel to the BBC and Russian, Chinese, and Japanese national television—has helped make the vision of human exploration of our neighbor world tangible to hundreds of millions of people around the globe.

The Mars Society holds its international convention every August. You can join either through our website at www.marssociety.org or by sending $50 ($25 for students) to Mars Society, 11111 W. 8th Ave., Unit A, Lakewood, CO 80215.

If you want to reach me, you can write in care of the Mars Society address above. If you want to help, sign up at the website so we can put you on the Mars Society electronic mailing list. If you join the Mars Society, you will also get access to our online library. You can get a fair number of my technical papers there, as well as those of many other authors, covering topics ranging from interplanetary propulsion technologies to the ethics of terraforming.

Finally, if you are not a joiner, then do what you can to spread the vision in your own way.

As the poet Percy Bysshe Shelley once said, "Poets are the unacknowledged legislators of the world."

Making history is not a spectator sport. It's your turn at the plate.

APPENDIX

FOUNDING DECLARATION OF THE MARS SOCIETY

The time has come for humanity to journey to Mars.

We're ready. Though Mars is distant, we are far better prepared today to send humans to Mars than we were to travel to the Moon at the commencement of the space age. Given the will, we could have our first teams on Mars within a decade.

The reasons for going to Mars are powerful.

We must go for the knowledge of Mars. Our robotic probes have revealed that Mars was once a warm and wet planet, suitable for hosting life's origin. But did it? A search for fossils on the Martian surface or microbes in groundwater below could provide the answer. If found, they would show that the origin of life is not unique to the Earth and, by implication, reveal a universe that is filled with life and probably intelligence as well. From the point of view of learning our true place in the universe, this would be the most important scientific enlightenment since Copernicus.

We must go for the knowledge of Earth. As we begin the twenty-first century, we have evidence that we are changing the Earth's atmosphere and environment in significant ways. It has become a critical matter for us to better understand all aspects of our environment. In this project, comparative planetology is a very powerful tool, a fact already shown by the role Venusian atmospheric studies played in our discovery of the potential threat of global warming by greenhouse gases. Mars, the planet most like Earth, will have even more to teach us about our home world. The knowledge we gain could be key to our survival.

We must go for the challenge. Civilizations, like people, thrive on challenge—and decay without it. The time is past for human societies to use war as a driving stress for technological progress. As the world moves toward unity, we must join together, not in mutual passivity, but in common enterprise, facing outward to embrace a greater and nobler challenge than that which we previously posed to each other. Pioneering Mars will provide such a challenge. Furthermore, a cooperative international exploration of Mars would serve as an example of how the same joint action could work on Earth in other ventures.

We must go for the youth. The spirit of youth demands adventure. A humans-to-Mars program would challenge young people everywhere to develop their minds to participate in the pioneering of a new world. If a Mars program were to inspire just a single extra percent of today's youth to scientific educations, the net result would be tens of millions more scientists, engineers, inventors, medical researchers, and doctors. These people will make innovations that create new industries, find new medical cures, increase income, and benefit the world in innumerable ways to provide a return that will utterly dwarf the expenditures of the Mars program.

We must go for the opportunity. The settling of the Martian New World is an opportunity for a noble experiment in which humanity has another chance to shed old baggage and begin the world anew, carrying forward as much of the best of our heritage as possible and leaving the worst behind. Such chances do not come often and are not to be disdained lightly.

We must go for our humanity. Human beings are more than merely another kind of animal—we are life's messenger. Alone of the creatures of the Earth, we have the ability to continue the work of creation by bringing life to Mars, and Mars to life. In doing so, we shall make a profound statement as to the precious worth of the human race and every member of it.

We must go for the future. Mars is not just a scientific curiosity; it is a world with a surface area equal to all the continents of Earth combined, possessing all the elements that are needed to support not only life, but technological society. It is a New World, filled with history waiting to be made by a new and youthful branch of human civilization that is waiting to be born. We must go to Mars to make that potential a reality. We must go, not for us, but for a people who are yet to be. We must do it for the Martians.

Believing therefore that the exploration and settlement of Mars is one of the greatest human endeavors possible in our time, we have gathered to found this Mars Society, understanding that even the best ideas for human action are never inevitable, but must be planned, advocated, and achieved by hard work. We call upon all other individuals and organizations of like-minded people to join with us in furthering this great enterprise. No nobler cause has ever been. We shall not rest until it succeeds.

The above declaration was signed and ratified by the seven hundred attendees at the Founding Convention of the Mars Society, held August 13–16, 1998, at the University of Colorado at Boulder, Colorado. If you agree, I invite you to join. Further information is available at www.marssociety.org or by writing the Mars Society, 11111 W. 8th Ave., Unit A, Lakewood, CO 80215.

GLOSSARY

ΔV: *See* delta-V.

aerobraking: A spacecraft maneuver using friction with a planetary atmosphere to decelerate from an interplanetary orbit to one about a planet.

aeroshell: A heat shield used to protect a spacecraft from atmospheric heating during aerobraking.

apogee: The highest point in an orbit about a planet.

atmospheric pressure: The pressure an atmosphere exerts. On Earth at sea level, the atmospheric pressure is 14.7 pounds per square inch. This amount of pressure is therefore known as one "atmosphere" or one "bar."

bipropellant: A rocket propellant combination including both a fuel and oxidizer. Examples include methane/oxygen, hydrogen/oxygen, kerosene/hydrogen peroxide, and so on.

BFR: SpaceX's concept for a reusable two-stage-to-orbit launch vehicle with a payload capacity of about 150 tons to low Earth orbit. The "Big F . . . ing Rocket" was originally introduced by Elon Musk as the "Interplanetary Transport System" (ITS) in September 2016, and renamed BFR in 2017, under which title it became widely known as discussed. It was then renamed "Starship" in November 2018. In this book we will call the former BFR a "Starship," while we call craft that travel to the stars "interstellar spaceships."

buffer gas: An effectively inert gas that is used to dilute the oxygen required to support breathing or combustion. On Earth, the 80 percent nitrogen found in air serves as a buffer gas.

cosmic ray: A particle, such as an atomic nucleus, traveling through space at very high velocity. Cosmic rays originate outside of our solar system. They typically have energies of billions of electron volts and require meters of solid shielding to stop.

cryogenic: Ultra cold. Liquid oxygen and hydrogen are both cryogenic fluids, as they require temperatures of –180° and –250°C, respectively, for storage.

delta-V: The velocity change required to move a spacecraft from one orbit to another. A typical delta-V (also written ΔV) required to go from low Earth orbit to a trans-Mars trajectory would be about 4 km/s.

departure velocity: The velocity of a spacecraft relative to a planet after effectively leaving the planet's gravitational field. Also known as hyperbolic velocity.

direct entry: A maneuver in which a spacecraft enters a planet's atmosphere and uses it to decelerate and land without going into orbit.

direct launch: A maneuver in which a spacecraft is launched directly from one planet to another without being assembled in orbit.

electrolysis: The use of electricity to split a chemical compound into its elemental components. Electrolysis of water splits it into hydrogen and oxygen.

endothermic: A chemical reaction requiring the addition of energy to occur.

equilibrium constant: A number that characterizes the degree to which a chemical reaction will proceed to completion. A very high equilibrium constant implies near-complete reaction.

ERV: Earth Return Vehicle. A spacecraft designed to return to Earth.

EVA: Extravehicular activity, meaning leaving your spacecraft to go outside in a space suit.

exhaust velocity: The speed of the gases emitted from a rocket nozzle.

exothermic: A chemical reaction that releases energy when it occurs.

fairing: The protective streamlined shell containing a payload that sits on top of a launch vehicle.

Falcon: The Falcons are a line of partially reusable launch vehicles developed and operated by the SpaceX company. The Falcon 9 can lift twenty-three tons to low Earth orbit. The Falcon Heavy can lift sixty-two tons to low Earth orbit.

free return trajectory: A trajectory that, after departing Earth, will eventually return to the Earth without any additional propulsive maneuvers.

geothermal energy: Energy produced by using naturally hot underground materials to heat a fluid, which can then be expanded in a turbine generator to produce electricity.

gigawatt: One billion watts. Equal to one thousand megawatts. Abbreviated GW.

gravity assist: A maneuver in which a spacecraft flying by a planet uses that planet's gravity to create a slingshot effect, which adds to the spacecraft's velocity without any requirement for the use of rocket propellant.

heliocentric: Centered about the sun. By a heliocentric orbit, what is meant is an orbit that transverses interplanetary space and is not bound to the Earth or any other planet.

Hohmann transfer orbit: An elliptical orbit, one of whose ends is tangent to the orbit of the planet of departure and whose other end is tangent to the orbit of the planet of destination. The Hohmann transfer orbit is the purest incarnation of the conjunction-class orbit and, as such, is the lowest energy path from one planet to another.

hydrazine: A rocket propellant whose formula is N_2H_4. Hydrazine is a monopropellant, which means that it can release energy by decomposing, without any additional oxidizer required for combustion.

hyperbolic velocity: The velocity of a spacecraft relative to a planet before entering, or after effectively leaving, the planet's gravitational field. Also known as approach or departure velocity.

hypersonic: A speed many times the speed of sound; in common usage, Mach 5 or greater.

ionosphere: The upper layer of a planet's atmosphere, in which a significant fraction of the gas atoms have split into free positively charged ions and negatively charged electrons. Because of the presence of freely moving charged particles, an ionosphere can reflect radio waves.

Isp: A commonly used abbreviation for specific impulse (*see* specific impulse).

ISPP: In situ propellant production. Making propellant on an extraterrestrial body out of local materials.

JSC: Johnson Space Center.

JPL: Jet Propulsion Lab.

Kelvin degrees: The Kelvin or "absolute" scale is a method of measuring temperature that starts with its zero point set at "absolute zero," the temperature at which a body in fact possesses no heat. The temperature 273 kelvin is the same as 0°C, the freezing point of water. Each additional degree kelvin corresponds to one additional degree Celsius.

kHz: Kilohertz, a measure of frequency used in radio. One kHz equals one thousand cycles per second.

kb/s: Kilobits per second.

km/s: Kilometers per second.

kW: Kilowatts.

kWe: Kilowatts of electricity.

kWe-hr: The total amount of energy provided by one kilowatt of electricity for one hour.

kWh: The total amount of energy associated with the use of one kilowatt for one hour.

m/s: Meters per second.

LEO: Low Earth orbit.

LOR: Lunar Orbit Rendezvous.

LOX: Liquid oxygen.

magsail: A magnetic sail (*see* magnetic sail).

magnetic sail: A device for propelling spacecraft using the pushing force of plasma on a magnetic field.

MAV: Mars Ascent Vehicle.

methanation reaction: A chemical reaction forming methane. In the Mars Direct mission, the methanation reaction is the Sabatier reaction, in which hydrogen is combined with carbon dioxide to produce methane and water.

millirem: One-thousandth of a rem (*see* rem).

minimum energy trajectory: The trajectory between two planets requiring the least amount of rocket propellant to attain (*see* Hohmann transfer orbit).

MOR: Mars Orbit Rendezvous.

MSR: Mars Sample Return.

MSR-ISPP: Mars Sample Return employing in situ propellant production.

MHz: Megahertz, a measure of frequency used in radio. One MHz equals one million cycles per second.

MWt: Megawatts of heat. One megawatt equals one thousand kilowatts.

MWe: Megawatts of electricity.

NEP: Nuclear Electric Propulsion. NEP systems use electricity from a fission power system to accelerate ions to very high speeds, thereby producing thrust.

NIMF: Nuclear rocket using indigenous Martian fuel.

NTR: Nuclear thermal rocket. NTRs use solid-core fission reactors to heat propellant gases (typically hydrogen, but methane, water, ammonia, and CO_2 are options) to very high temperatures, thereby producing thrust.

perigee: The lowest point in an orbit around a planet.

pyrolyze: The use of heat to split a compound into its elemental constituents.

regolith: What most commonly refer to as dirt.

rem: The measure of radiation dose most commonly used in the United States. One hundred rem equals one Sievert, the European unit. It is estimated that radiation doses of about sixty to eighty rem are sufficient to increase a person's probability of fatal cancer at some time later in life by 1 percent. Typical background radiation on Earth is about 0.2 rem/year.

RWGS: Reverse water-gas shift reaction.

RTG: Radioisotope thermoelectric generator.

Sabatier reaction: A reaction in which hydrogen and carbon dioxide are combined to produce methane and water. The Sabatier reaction is exothermic, with a high equilibrium constant (*see* equilibrium; exothermic).

Saturn V: The heavy-lift launch vehicle used to send the Apollo astronauts to the moon. The Saturn V could lift about 140 tons to LEO.

SEI: Space Exploration Initiative.

SNC meteorites: Named for the locations where the first three were found (Shergotty, Nakhla, and Chassigny), SNC meteorites are believed on the basis of very strong chemical, geologic, and isotopic evidence to be debris thrown off of Mars by impacting meteorites.

sol: One Martian day.

solar flare: A sudden eruption on the surface of the sun that can deliver immense amounts of radiation across vast stretches of space.

solar sail: A device for propelling a spacecraft by utilizing the pushing force of sunlight.

specific impulse: The specific impulse of a rocket engine is the number of seconds it can make a pound of propellant deliver a pound of thrust. If you multiply the specific impulse of a rocket engine, given in seconds, by 9.8, you will obtain the engine's exhaust velocity in units of meters/second. Specific impulse is generally viewed as the most important factor in judging a rocket engine's performance. Frequently abbreviated "Isp."

SSTO: Single-stage-to-orbit.

Starship: A fully reusable two-stage-to-orbit launch vehicle with a payload capacity of 150 tons being developed by the SpaceX company. Formerly known as the BFR.

telerobotic operation: Remote control of some device, such as a small Mars rover equipped with video cameras, by human operators at a significant distance away.

ton: In this book, a ton is a metric ton, or one thousand kilograms. Equal to 2,200 pounds.

thrust: The amount of force a rocket engine can exert to accelerate a spacecraft.

TMI: Trans-Mars injection, a maneuver that places a spacecraft on a trajectory to Mars.

TW: Terawatt; one terawatt equals one million megawatts. Human civilization today uses about 23 TW.

TW-year: The total amount of energy associated with the use of one terawatt for one year.

tokamak: An experimental fusion power machine employing a toroidal magnetic field and vacuum chamber to confine a high-temperature plasma.

W/kg: watts per kilogram.

vapor pressure: The pressure exerted by the gas emitted by a substance at a certain temperature. At 100°C, the vapor pressure of water is greater than the Earth's atmospheric pressure, and so it will boil.

NOTES

INTRODUCTION

1. Nikolai Kardashev, "Transmission of Information by Extraterrestrial Civilizations," *Soviet Astronomy* 8 (1964): 217; Nikolai Kardashev, "On the Inevitability and the Possible Structures of Supercivilizations," in *The Search for Extraterrestrial Life: Recent Developments: Proceedings of the Symposium, Boston, MA, June 18–21, 1984* (Dordrecht, Netherlands: D. Reidel, 1985), pp. 497–50; I. S. Shklovskii and Carl Sagan, *Intelligent Life in the Universe* (New York: Dell, 1966). The original Kardashev scheme defined a Type I civilization as one that used all the energy falling on its planet, a Type II civilization as one using all the energy of its star, and a Type III civilization as one using all the energy of all the stars in its galaxy. I don't believe that this particular form of such a scale is useful, as no civilization will ever use all the sunlight falling on its home planet's surface and thereby become Type I, let alone a civilization of the higher types. Nevertheless, Kardashev's effort to create a classification system for advanced spacefaring civilization was an important step. So I have adapted his general schema to what I consider more a useful metric for measuring civilizational progress as presented here.

CHAPTER 1. BREAKING THE BONDS OF EARTH

1. Julian Guthrie. *How to Make a Spaceship: A Band of Renegades, an Epic Race, and the Birth of Private Spaceflight* (New York: Penguin, 2016).
2. Robert Zubrin with Richard Wagner, *The Case for Mars: The Plan to Settle the Red Planet, and Why We Must*, 2nd ed. (New York: Free Press, 2011).
3. Wikipedia, s.v. "Mars Gravity Biosatellite," last modified October 26, 2018, 14:23, https://en.wikipedia.org/wiki/Mars_Gravity_Biosatellite; "Mars Society Launches Translife Mission," *Spaceref*, August 30, 2001, http://www.spaceref.com/news/viewpr.html?pid=5881 (accessed October

14, 2018); "Translife Mission Experiment Sees Mice Born at 25 RPM," *Space Daily*, October 15, 2001, http://www.spacedaily.com/news/mars-base-01f .html (accessed October 12, 2018).

4. Aaron Rowe, "SpaceX Did It: Falcon 1 Made It to Space, " *Wired*, September 28, 2008, https://www.wired.com/2008/09/space-x-did-it/ (accessed October 14, 2018).

5. Kenneth Chang, "Private Rocket Has Successful First Flight," *New York Times*, June 4, 2010, https://www.nytimes.com/2010/06/05/science/ space/05rocket.html (accessed October 14, 2018).

6. Adam Mann, "Private Plan to Send Humans to Mars in 2018 Might Not Be So Crazy," *Wired*, February 27, 2013, https://www.wired .com/2013/02/inspiration-mars-foundation/ (accessed October 14, 2014).

7. Gerard K. O'Neill, *The High Frontier: Human Colonies in Space* (New York: William Morrow, 1976).

8. Nicholas St. Fleur, "Jeff Bezos Says He Is Selling $1 Billion a Year in Amazon Stock to Finance Race to Space," *New York Times*, April 5, 2017, https://www.nytimes.com/2017/04/05/science/blue-origin-rocket-jeff-bezos-amazon-stock.html (accessed October 14, 2018).

9. It is interesting to note that an innovative American-founded space launch start-up called Firefly has chosen to locate its research and development branch in Ukraine rather than in Russia. While both nations have inherited significant parts of the Soviet Union's space technology, relatively free Ukraine's smaller portion is far more investible than that of kleptocratic Russia.

CHAPTER 2. FREE SPACE

1. James Titcomb, "Elon Musk Plans London to New York Flights in 29 Minutes," *Telegraph*, September 29, 2017, https://www.telegraph.co.uk/ technology/2017/09/29/elon-musk-unveils-plans-london-new-york -rocket-flights-30-minutes/ (accessed October 14, 2018). Note the BFR, "Big F . . . ing Rocket," was originally introduced by Musk as the Interplanetary Transport System (ITS) in September 2016 and renamed BFR in early 2017, under which title it became widely known and discussed. It was then renamed Starship in November 2018. In this book, we will call the former BFR a Starship, while craft that engage in trips to the stars will be called "interstellar spaceships."

2. Christian Davenport, "Richard Branson's Virgin Galactic Just

Got Another Step Closer to Flying Tourists to Space," *Washington Post*, May 29, 2018, https://www.washingtonpost.com/news/the-switch/wp/2018/05/29/richard-bransons-virgin-galactic-just-got-another-step-closer-to-flying-tourists-to-space/?noredirect=on&utm_term=.e548415f3697 (accessed October 14, 2018).

3. Stefanie Waldek, "How to Become a Space Tourist: 8 Companies (Almost) Ready to Launch," *Popular Science*, April 20, 2018, https://www.popsci.com/how-to-become-a-space-tourist (accessed October 14, 2018).

4. Julissa Treviño, "A Space Hotel Could Be Coming Soon to Skies near You: Bigelow Aerospace Wants to Launch Two Inflatable Modules for a Space Habit as Early as 2021," *Smithsonian*, March 1, 2018, https://www.smithsonianmag.com/smart-news/space-hotel-it-could-happen-near-future-180968248/ (accessed October 14, 2018). For a comical look at the possible downside of space hotel vacations, seek out Diana Gallagher's "A Reconsideration of Anatomical Docking Maneuvers in a Zero-G Environment"—one recording is at "Free Fall & Other Delights 11—A Reconsideration (Zero-G Sex)," YouTube video, 4:12, posted by weyrdmusicman, August 18, 2011, https://www.youtube.com/watch?v=QrC6paNUry0.

5. Stan Schroeder, "In 5 Years, the Average American Will Use 22GB of Mobile Data Per Month, Report Says," *Mashable*, June 1, 2016, https://mashable.com/2016/06/01/ericsson-mobility-report-2016/#ZziDG2D80sqH (accessed October 14, 2016).

6. Peter B. de Selding, "WorldVu, a Satellite Startup Aiming to Provide Global Internet Connectivity, Continues to Grow Absent Clear Google Relationship," *Space News*, September 3, 2014, https://spacenews.com/41755worldvu-a-satellite-startup-aiming-to-provide-global-internet/ (accessed October 14, 2018).

7. Caleb Henry, "OneWeb Shifts First Launch to Year's End," *Space News*, May 1, 2018, https://spacenews.com/oneweb-shifts-first-launch-to-years-end/ (accessed October 14, 2018).

8. David Grossman, "The Race for Space-Based Internet Is On," *Popular Mechanics*, January 3, 2018, https://www.popularmechanics.com/technology/infrastructure/a14539476/the-race-for-space-based-internet-is-on/ (accessed October 14, 2018).

9. Patrick Daniels, "SpaceX Starlink: Here's Everything You Need to Know," *Digital Trends*, August 5, 2018, https://www.digitaltrends.com/cool-tech/spacex-starlink-elon-musk-news/ (accessed October 14, 2018).

10. Committee on Achieving Science Goals with CubeSats, *Achieving*

Science with CubeSats: Thinking Inside the Box (Washington, DC: National Academies Press, 2016).

    11. Sandra Erwin, "US Intelligence: Russia and China Will Have 'Operational' Anti-Satellite Weapons in a Few Years," *Space News*, February 14, 2018, https://spacenews.com/u-s-intelligence-russia-and-china-will -have-operational-anti-satellite-weapons-in-a-few-years/ (accessed October 14, 2018).

    12. John T. Correll, "Targeting the Luftwaffe," *Air Force Magazine*, March 2018, http://www.airforcemag.com/MagazineArchive/Pages/2018/ March%202018/Targeting-the-Luftwaffe.aspx (accessed October 14, 2018).

CHAPTER 3. HOW TO BUILD A LUNAR BASE

    1. Robert Zubrin, "Cancel the Lunar Orbit Tollbooth," *National Review*, September 13, 2018, https://www.nationalreview.com/2018/09/nasa -lunar-orbiting-platform-gateway-should-be-canceled/ (accessed October 14, 2014).

    2. Wendell Mendell, *Lunar Bases and Space Activities of the 21st Century* (Houston: Lunar and Planetary Institute, 1995).

    3. L. Haskin and P. Warren, "Lunar Chemistry," in *Lunar Sourcebook*, ed. G. Heiken, D. Vaniman, and B. French (Cambridge: Cambridge University Press, 1991), chap. 8.

    4. Leonard David, "Beyond the Shadow of a Doubt, Water Ice Exists on the Moon," *Scientific American*, August 21, 2018, https://www.scientific american.com/article/beyond-the-shadow-of-a-doubt-water-ice-exists-on -the-moon/ (accessed October 14, 2018).

    5. Robert Zubrin, "Moon Direct," *New Atlantis*, Summer/Fall 2018, https://www.thenewatlantis.com/publications/moon-direct (accessed January 20, 2019).

    6. Apollo News Reference, "Lunar Module," NASA, https://www.hq .nasa.gov/alsj/LM04_Lunar_Module_ppLV1-17.pdf (accessed October 14, 2018).

    7. Harrison Schmitt, *Return to the Moon: Exploration, Enterprise, and Energy in the Human Settlement of Space* (New York: Copernicus Books, 2006).

    8. Andy Coghlan, "Technology: Moon Rocks May Help Colonisers Breathe Easy," *New Scientist*, April 18, 1992, https://www.newscientist .com/article/mg13418173-900-technology-moon-rocks-may-help -colonisers-breathe-easy/ (accessed October 14, 2018).

9. Y. Artsutanov, "V kosmos na elektrovoze" [Into Space by Funicular Railway], *Komsomolskaya pravda*, July 31, 1960. Contents described in Lvov, *Science* 158 (November 17, 1967): 946.

10. J. Pearson, "The Orbital Tower: A Spacecraft Launcher Using the Earth's Rotational Energy," *Acta Astronautica* 2 (1974): 785–99.

11. For the required equations, see Robert Zubrin, "The Hypersonic Skyhook," *Journal of the British Interplanetary Society* 48 (March 1995): 123–25. A summary treatment can also be found in Stan Schmidt and Robert Zubrin, *Islands in the Sky: Bold New Ideas for Colonizing Space* (New York: Wiley, 1996), chap. 2. The calculation that the tether would have a mass less than twice its payload assumes a tapered tether with a tensile strength of 2,000 megapascals. For comparison, Kevlar has a yield strength of 2,800 MPa, Spectra of 3,200 MPa, and Dyneema of 3,500 MPa.

CHAPTER 4. MARS

1. Mike Carr, *The Surface of Mars* (New Haven, CT: Yale University Press, 1981).

2. Robert Zubrin with Richard Wagner, *The Case for Mars: The Plan to Settle the Red Planet, and Why We Must*, 2nd ed. (New York: Free Press, 2011).

3. Katherine Hignett, "Water on Mars: Huge Lake Detected below Red Planet Surface in 'Major Milestone' Discovery," *Newsweek*, July 25, 2018, https://www.newsweek.com/water-mars-huge-lake-detected-below-red -planets-surface-major-milestone-1041265 (accessed October 14, 2018).

4. N. E. Putzig et al., "Subsurface Structure of Planum Boreum from Mars Reconnaissance Orbiter Shallow Radar Soundings," *Icarus* 204, no. 2 (2009): 443–57; N. B. Karlsson, L. S. Schmidt, and C. S. Hvidberg, "Volume of Martian Midlatitude Glaciers from Radar Observations and Ice Flow Modeling," *Geophysical Research Letters* 42, no. 8 (March 18, 2015): 2627–33, https://agupubs.onlinelibrary.wiley.com/doi/full/10.1002/ 2015GL063219 (accessed October 14, 2018); Jet Propulsion Lab, "Steep Slopes on Mars Reveal Structure of Buried Ice," press release, January 11, 2018, https://www.jpl.nasa.gov/news/news.php?feature=7038 (accessed October 14, 2018).

5. Kenneth Chang, "Falcon Heavy, in a Roar of Thunder, Carries SpaceX's Ambition into Orbit," *New York Times*, February 6, 2018, https:// www.nytimes.com/2018/02/06/science/falcon-heavy-spacex-launch.html (accessed October 14, 2018).

6. Noah Robischon and Elizabeth Segran, "Elon Musk's Mars Mission Revealed: SpaceX's Interplanetary Transport System," *Fast Company*, September 27, 2016, https://www.fastcompany.com/3064139/elon-musks-mars-mission-revealed-spacexs-interplanetary-transport-system (accessed October 14, 2018).

7. Robert Zubrin, "Colonizing Mars: A Critique of the SpaceX Interplanetary Transport System," *New Atlantis*, October 21, 2016, https://www.thenewatlantis.com/publications/colonizing-mars (accessed October 14, 2018).

8. Adam Baidawi and Kenneth Chang, "Elon Musk's Mars Vision: A One-Size-Fits-All Rocket. A Very Big One," *New York Times*, September 27, 2017, https://www.nytimes.com/2017/09/28/science/elon-musk-mars.html (accessed October 14, 2018).

9. Carol Stoker and Carter Emmart, *Strategies for Mars* (San Diego: Univelt, 1996); Robert Zubrin, *From Imagination to Reality: Mars Exploration Studies of the Journal of the British Interplanetary Society* (San Diego: Univelt, 1997).

10. Zubrin, *Case for Mars*, chap. 9; Robert Zubrin and Christopher McKay, "Technological Requirements for Terraforming Mars," *AIAA* (1993), http://citeseerx.ist.psu.edu/viewdoc/download?doi=10.1.1.24.8928&rep=rep1&type=pdf (accessed October 14, 2018).

11. In August 2018, a much-publicized paper by University of Colorado professor Bruce Jakowski, the principal investigator of the MAVEN orbiter, asserted that there could not be enough CO_2 on Mars to create an appreciable atmosphere, because on the basis of MAVEN measurements, Mars has lost about 7 psi of pressure to space over the past four billion years. In fact, these results show the exact opposite, since Mars would have needed more than 20 psi of CO_2 in its atmosphere four billion years ago to have been warm enough for liquid water. So plenty of CO_2 must still be soaked in the soil. Mike Brown, "Elon Musk Wants to Terraform Mars, and He's Refusing to Back Down," *Yahoo News*, August 1, 2018, https://www.yahoo.com/news/elon-musk-wants-terraform-mars-123100942.html (accessed October 15, 2018).

CHAPTER 5. ASTEROIDS FOR FUN AND PROFIT

1. John Lewis, *Mining the Sky: Untold Riches from the Asteroids, Comets, and Planets* (New York: Helix Books, 1996).

2. David Harland, *Jupiter Odyssey: The Story of NASA's Galileo Mission* (Chichester, UK: Praxis, 2000).

3. Jim Bell and Jaqueline Mitton, *Asteroid Rendezvous: NEAR Shoemaker's Adventures at Eros* (Cambridge: Cambridge University Press, 2002).

4. Mika McKinnon, "Everything That Could Go Wrong for Hayabusa Did, and Yet It Still Succeeded," *Gizmodo*, October 15, 2015, https://gizmodo .com/everything-that-could-go-wrong-for-hayabusa-did-and-ye -1730940605 (accessed October 16, 2018).

5. Tim Sharp, "Rosetta Spacecraft: To Catch a Comet," *Space*, February 27, 2017, https://www.space.com/24292-rosetta-spacecraft.html (accessed October 16, 2018).

6. Mike Wall, "The End Is Near for NASA's Historic Dawn Mission to the Asteroid Belt," *Space*, September 7, 2018, https://www.space.com/ 41759-nasa-dawn-mission-asteroid-belt-nearly-over.html (accessed October 16, 2018).

7. Mike Wall, "Success! Hopping, Shoebox-Sized Lander Touches Down Safely on Asteroid Ryugu," *Space*, October 3, 2018, https://www.space .com/42003-hayabusa2-drops-mascot-lander-on-asteroid-ryugu.html (accessed October 16, 2018).

8. Amber Jorgensen, "OSIRIS-REx Snaps Its First Pic of Asteroid Bennu," *Astronomy*, August 28, 2018, http://www.astronomy.com/ news/2018/08/osiris-rex-snaps-its-first-pic-of-asteroid-bennu (accessed October 16, 2018).

9. John Lewis and Ruth Lewis, *Space Resources: Breaking the Bonds of Earth* (New York: Columbia University Press, 1987).

10. Robert Zubrin with Richard Wagner, *The Case for Mars: The Plan to Settle the Red Planet, and Why We Must*, 2nd ed. (New York: Free Press, 2011).

11. Charles Cockell, *Extra-Terrestrial Liberty: An Enquiry into the Nature and Causes of Tyrannical Government beyond the Earth* (Edinburgh: Shoving Leopard, 2013).

12. Maia Weinstock, "Oxygen-Creating Instrument Selected to Fly on the Upcoming Mars 2020 Mission," *Phys.org*, August 1, 2014, https://phys .org/news/2014-08-oxygen-creating-instrument-upcoming-mars-mission .html (accessed October 16, 2018).

13. Lewis and Lewis, *Space Resources*.

CHAPTER 6. THE OUTER WORLDS

1. W. Burrows, *Exploring Space: Voyages in the Solar System and Beyond* (New York: Random House, 1990).

2. David Harland, *Jupiter Odyssey: The Story of NASA's Galileo Mission* (Chichester, UK: Praxis, 2000).

3. Sarah Kaplan, "Ingredients for Life Discovered Gushing out of Saturn's Moon," *Washington Post*, June 27, 2018, https://www.washington post.com/news/speaking-of-science/wp/2018/06/27/complex-organic -molecules-discovered-in-enceladuss-plumes/?utm_term=.0466618a6f94 (accessed October 16, 2018).

4. Elizabeth Howell, "Europa Clipper: Sailing to Jupiter's Icy Moon," *Space*, July 21, 2017, https://www.space.com/37282-europa-clipper.html (accessed October 16, 2018).

5. Eric Berger, "The Billion-Dollar Question: How Does the Clipper Mission Get to Europa?" *Ars Technica*, April 16, 2018, https://arstechnica .com/science/2018/04/if-were-really-going-to-europa-nasa-needs-to-pick -a-rocket-soon/ (accessed January 20, 2019).

6. Kenneth Chang, "Back to Saturn? Five Missions Proposed to Follow Cassini," *New York Times*, September 15, 2017, https://www.nytimes.com/ 2017/09/15/science/saturn-cassini-return.html (accessed October 16, 2018).

7. Let the power of a rocket engine be given by P, thrust by T, propellant mass flow by M, and exhaust velocity by U. The $P=MU^2/2$, and $T=MU$. Using some algebra, we see that $U=\mathrm{sqrt}(2P/M)$, and $T=\mathrm{sqrt}(2PM)$. So thrust increases and exhaust velocity decreases in proportion to the square root of the propellant mass flow. Also, note that $T=2P/U$. So for a given amount of power, thrust *decreases* in proportion to the exhaust velocity. This is why, in a power limited system, you want *lower* exhaust velocity for shorter trips. If the exhaust velocity is too high, it takes too long to get the spacecraft up to speed. An ultrahigh-exhaust-velocity spacecraft is like a car that gets a thousand miles to the gallon but goes from zero to sixty in an hour. It's useless for driving around town but could be attractive for long trips on the highway. That's why for trips to the moon and Mars we want high-thrust, lower-exhaust-velocity engines like chemical rockets or nuclear thermal propulsion. But for trips to the outer solar system, high-exhaust -velocity, low-thrust systems like ion drives and fusion propulsion become advantageous and, for travel to the stars, essential.

8. R. Terra, "Islands in the Sky: Human Exploration and Settlement

of the Oort Cloud," in *Islands in the Sky*, ed. S. Schmidt and R. Zubrin (New York: Wiley, 1995); B. Finney and E. Jones, eds., *Interstellar Migration and the Human Experience* (Berkeley: University of California Press, 1985); Freeman Dyson, "Warm-Blooded Plants and Freeze-Dried Fish: When Emigration from Earth to a Planet or a Comet Becomes Cheap Enough for Ordinary People to Afford, People Will Emigrate," *Atlantic*, November 1997, https://www.theatlantic.com/past/docs/issues/97nov/space.htm (accessed October 16, 2018). For a fun speculation about a comet colonization mission envisioned by two top science fiction writers who are also working scientists, see Greg Benford and David Brin, *The Heart of the Comet* (New York: Bantam Spectra, 1986).

CHAPTER 7. REACHING FOR THE STARS

1. R. G. Ragsdale, "To Mars Is 30 Days by Gas Core Nuclear Rockets," *Astronautics and Aeronautics* 65 (January 1972); R. G. Ragsdale, *High Specific Impulse Gas Core Reactors*, NASA TM X-2243 (Cleveland: NASA Lewis Research Center, March 1971); T. Latham and C. Joyner, "Summary of Nuclear Light Bulb Development Status," AIAA 91-3512 (presented at the AIAA/NASA/OAI Conference on Advanced SEI Technologies, Cleveland, OH, September 1991).

2. A. Martin and A. Bond, "Nuclear Pulse Propulsion: A Historical Review of an Advanced Propulsion Concept," *Journal of the British Interplanetary Society* 32 (1979): 283–310.

3. R. Zubrin, "Nuclear Salt Water Rockets: High Thrust at 10,000 Sec Isp," *Journal of the British Interplanetary Society* 44 (1991): 371–76.

4. S. K. Borowski, "A Comparison of Fusion/Antiproton Propulsion Systems for Interplanetary Travel," AIAA 87-1814 (presented at the 23rd AIAA/ASME Joint Propulsion Conference, San Diego, CA, June 29 to July 2, 1987).

5. R. Forward and Joel Davis, *Mirror Matter: Pioneering Antimatter Physics* (New York: Wiley Science Editions, 1988).

6. L. Friedman, *Starsailing: Solar Sails and Interstellar Travel* (New York: John Wiley and Sons, 1988).

7. R. Forward, "Roundtrip Interstellar Travel Using Laser Pushed Lightsails," *Journal of Spacecraft and Rockets* 21 (1984): 187–95.

8. R. Bussard, "Galactic Matter and Interstellar Flight," *Acta Astronautica* 6 (1960): 179–96.

9. R. Zubrin and D. Andrews, "Magnetic Sails and Interplanetary Travel," AIAA 89-2441 (presented at the 25th AIAA/ASME Joint Propulsion Conference, Monterey, CA, July 10–12, 1989), reprinted in *Journal of Rockets and Spaceflight*, April 1991; R. Zubrin, "The Magnetic Sail," in *Islands in the Sky*, ed. S. Schmidt and R. Zubrin (New York: Wiley, 1995).

10. D. Andrews and R. Zubrin, "MagOrion" (presented at the 33rd AIAA/ASME Joint Propulsion Conference, Seattle, WA, July 7–9, 1997).

11. Pekka Janhunen, "Electric Sail for Space Craft Propulsion," *Journal of Propulsion* 20, no. 4 (2005): Technical Notes, 763–64.

12. Robert Zubrin, "Dipole Drive for Space Propulsion," *Journal of the British Interplanetary Society* 70, no. 12 (December 2017): 442–48.

13. Marc G. Millis, *Assessing Potential Propulsion Breakthroughs*, NASA/TM 2005-213-998 (Cleveland: Glenn Research Center, December 2005), https://ntrs.nasa.gov/archive/nasa/casi.ntrs.nasa.gov/20060000022.pdf (accessed November 25, 2018).

14. Harold White et al., "Measurement of Impulse Thrust from a Closed Resonant Frequency Cavity in Vacuum," *Journal of Propulsion and Power* 33, no. 4 (2017): 830–41, https://arc.aiaa.org/doi/10.2514/1.B36120 (accessed November 25, 2018); Mike Wall, "Impossible EM Drive May Really Be Impossible," *Space*, May 23, 2018, https://www.space.com/40682 -em-drive-impossible-space-thruster-test.html (accessed November 25, 2018); Brian Wang, "Mach Effect Propellantless Drive Gets NIAC Phase 2 and Progress Towards Great Interstellar Propulsion," *Next Big Future*, April 1, 2018, https://www.nextbigfuture.com/2018/04/mach-effect -propellantless-drive-gets-niac-phase-2-and-progress-to-great-interstellar -propulsion.html (accessed November 25, 2018).

15. Allan Boyle, "'Eggs' for Alien Earths? At 94, Physicist Freeman Dyson's Brain Is Still Going Strong," *Geekwire*, May 8, 2018, https://www .geekwire.com/2018/eggs-alien-earths-freeman-dyson/ (accessed October 22, 2018).

16. Karel Čapek, *War with the Newts* (London: Allen & Unwin, 1937; New York: Bantam, 1955). Čapek was a Czech humanist and antifascist. In this book, he brilliantly satirizes practically everything, from lawyers, newspapers, business, actors, and the Ku Klux Klan to the Western allies' policy of appeasing the Nazis. Interestingly, Čapek is also the person who invented the word "robot" in his 1920 science fiction play *RUR*.

CHAPTER 8. TERRAFORMING

1. James Oberg, *New Earths* (Harrisburg, PA: Stackpole Books, 1981); Martyn Fogg, *Terraforming: Engineering Planetary Environments* (Warrendale, PA: Society of Automotive Engineers, 1995); J. Pollack and C. Sagan, "Planetary Engineering," in *Resources of Near Earth Space*, ed. J. Lewis, M. Mathews, and M. Guerreri (Tucson: University of Arizona Press, 1993), pp. 921–50.

2. C. Sagan, "The Planet Venus," *Science* 133 (1961): 849–58.

3. J. Kasting, "Runaway and Moist Greenhouse Atmospheres and the Evolution of Earth and Venus," *Icarus* 74 (1988): 472–94.

4. "Climate Change Indicators: Length of Growing Season," United States Environmental Protection Agency, https://www.epa.gov/climate -indicators/climate-change-indicators-length-growing-season (accessed November 17, 2018).

5. "Climate Change Indicators: US and Global Precipitation," United States Environmental Protection Agency, https://www.epa.gov/climate -indicators/climate-change-indicators-us-and-global-precipitation (accessed November 17, 2018).

6. "Carbon Dioxide Fertilization Greening the Earth, Study Finds," NASA, https://www.nasa.gov/feature/goddard/2016/carbon-dioxide -fertilization-greening-earth (accessed November 17, 2018).

7. Russ George, "We Can Bring Back Healthy Fish in Abundance Almost Everywhere," personal website, http://russgeorge. net/2014/04/11/bring-back-fish-everywhere/ accessed November 17, 2018).

8. NASA Goddard Spaceflight Center, http://disc.sci.gsfc.nasa.gov/ giovanni/giovanni_user_images#iron_bloom_northPac (accessed November 17, 2018).

9. Naomi Klein, *This Changes Everything: Capitalism against the Climate* (New York: Simon and Schuster, 2014); Naomi Klein, "Geoengineering: Testing the Waters," *New York Times*, October 28, 2012, http://www.nytimes.com/2012/10/28/opinion/sunday/geoengineering -testing-the-waters.html?pagewanted=all (accessed January 25, 2019).

10. Margaret Munro, "Ocean Fertilization: 'Rogue Climate Hacker' Russ George Raises Storm of Controversy," *Vancouver Sun*, October 18, 2012.

11. Assuming 1 percent efficiency of phytoplankton in converting CO_2 into biomass, and an average day/night sunlight flux of two hundred watts per square meter, it would take ten million square kilometers, or

3 percent of the oceans' total area, to capture all the CO_2 emissions from humanity's current twenty-terawatt fossil fuel combustion. Or, put another way, humanity currently burns ten billion tons of carbon per year, while the ocean captures one hundred billion tons. So a net increase of 10 percent of total ocean productivity, which can be accomplished by raising a much smaller fraction of the oceans' desert expanses to match the productivity of its most fertile regions, would suffice to capture all human CO_2 emissions. If done in optimal fashion, up to a billion tons of food could be produced in the process, enough to provide about a pound per day to every person on Earth.

12. K. Eric Drexler, *Engines of Creation* (New York: Anchor Books/ Doubleday, 1987).

13. Olaf Stapledon, *Star Maker* (London: Methuen, 1937; New York: Dover, 1968).

14. Elizabeth Tasker, *The Planet Factory: Exoplanets and the Search for a Second Earth* (New York: Bloomsbury Sigma, 2017).

CHAPTER 9. FOR THE KNOWLEDGE

1. Elizabeth Howell, "Kepler Space Telescope: Exoplanet Hunter," *Space*, March 26, 2018, https://www.space.com/24903-kepler-space -telescope.html (accessed October 22, 2018).

2. John Wenz, "TESS Space Telescope Will Find Thousands of Planets, but Astronomers Seek a Select Few," *Smithsonian*, September 26, 2018, https://www.smithsonianmag.com/science-nature/tess-space-telescope -will-find-thousands-planets-astronomers-seek-select-few-180970411/ (accessed October 22, 2018); Elizabeth Howell, "NASA's James Webb Space Telescope: Hubble's Cosmic Successor," *Space*, July 17, 2018, https:// www.space.com/21925-james-webb-space-telescope-jwst.html (accessed October 22, 2018).

3. Alison Klesman, "Where Does WFIRST Stand Now?" *Astronomy*, May 31, 2018, http://www.astronomy.com/news/2018/05/where-does-wfirst -stand-now (accessed October 22, 2018). For a more technical presentation, see N. Gehrels and D. Spergel, "Wide-Field InfraRed Survey Telescope (WFIRST) Mission and Synergies with LISA and LIGO-Virgo," https://arxiv. org/ftp/arxiv/papers/1411/1411.0313.pdf (accessed October 22, 2018).

4. "The Gravitational Universe: The Science Case for LISA," LISA team website, https://www.lisamission.org/articles/lisa-mission/gravitational -universe-science-case-lisa (accessed October 22, 2018).

5. Yuen Yiu, "Scientists Reveal Plans for Future Experiments to Study the Faint Remnants Left Behind by the Big Bang," *Inside Science*, February 23, 2017, https://www.insidescience.org/news/looking-deeper-our-cosmic-past (accessed October 22, 2018).

6. M. Meixner et al., "The Far-Infrared Surveyor Mission Study: Paper I, the Genesis," Space Telescope Science Institute, https://arxiv.org/ftp/arxiv/papers/1608/1608.03909.pdf (accessed October 22, 2018).

7. Feryal Özel and Alexey Vikhlinin, "Lynx Interim Report," Smithsonian Astrophysical Observatory, August 2018, https://wwwastro.msfc.nasa.gov/lynx/docs/LynxInterimReport.pdf (accessed October 22, 2018).

8. Bertrand Manneson et al., "The Habitable Exoplanet (HabEx) Imaging Mission: Preliminary Science Drivers and Technical Requirements," in *Space Telescopes and Instrumentation 2016: Optical, Infrared, and Millimeter Wave*, Proceedings of SPIE, vol. 9904, ed. Howard A. MacEwen et al. (Bellingham, WA: SPIE, 2016), https://pdfs.semanticscholar.org/6516/a74971950fa0efbcfd8f3e02edd850a687f3.pdf (accessed October 22, 2018).

9. "Origins Space Telescope," NASA, https://asd.gsfc.nasa.gov/firs/ (accessed October 22, 2018).

10. Ethan Siegel, "New Space Telescope, 40 Times the Power of Hubble, to Unlock Astronomy's Future," *Forbes*, September 19, 2017, https://www.forbes.com/sites/startswithabang/2017/09/19/new-space-telescope-40-times-the-power-of-hubble-to-unlock-astronomys-future/#7d9ccebe7e81 (accessed October 22, 2018).

11. S. Arrhenius, *Worlds in the Making: The Evolution of the Universe* (New York and London: Harper Brothers, 1908).

12. R. Zubrin, "Detection of Extraterrestrial Civilizations via the Spectral Signature of Advanced Interstellar Spacecraft," in *Progress in the Search for Extraterrestrial Life: 1993 Bioastronomy Symposium*, Seth Shostak, ed. (San Francisco: Astronomical Society of the Pacific, 1996), pp. 487–96.

13. Seth Shostak, *Sharing the Universe* (Berkeley, CA: Berkeley Hills Books, 1998).

14. W. Herkewitz, "Scientists Turn Bacteria into Living Hard Drives," *Popular Science*, June 9, 2016; R. Service, "DNA Could Store All the World's Data in One Room," *Science*, March 2, 2017.

15. Robert Zubrin, "Interstellar Panspermia Reconsidered," *Journal of the British Interplanetary Society* 54 (2001): 262–69.

16. Robert Zubrin, "Interstellar Communications Using Microbial Data Storage: Implications for SETI," *Journal of the British Interplanetary Society* 70, no. 5/6 (May/June 2017).

17. Paul Davies, *The Goldilocks Enigma: Why Is the Universe Just Right for Life?* (New York: Mariner Books, 2008).

18. John Barrow and Frank Tipler, *The Anthropic Cosmological Principle* (New York: Oxford University Press, 1988).

19. Lee Smolin, *The Life of the Cosmos* (New York: Oxford University Press, 1997).

20. Stuart Kauffman, *At Home in the Universe: The Search for the Laws of Self-Organization and Complexity* (New York: Oxford University Press, 1996).

21. I. S. Shklovskii and Carl Sagan, *Intelligent Life in the Universe* (New York: Delta Books, 1966).

CHAPTER 10. FOR THE CHALLENGE

1. Christopher Stringer and Robin McKie, *African Exodus: The Origins of Modern Humanity* (New York: Henry Holt, 1997).

2. James Shreve, *The Neanderthal Enigma: Solving the Mystery of Modern Human Origins* (New York: Avon Books, 1995).

3. William McNeill, *The Rise of the West* (New York: Mentor Books, 1965).

4. Carroll Quigley, *The Evolution of Civilizations* (Indianapolis, IN: Liberty Fund, 1961).

5. Francis Fukuyama, *The End of History* (New York: Free Press, 1992).

6. James Horgan, *The End of Science* (New York: Broadway Books, 1997).

7. Thomas D. Snyder, *120 Years of American Education: A Statistical Portrait* (Washington, DC: US Department of Education, 1993), pp. 85–87, https://nces.ed.gov/pubs93/93442.pdf (accessed November 24, 2018).

CHAPTER 11. FOR OUR SURVIVAL

1. Carl Sagan, *Pale Blue Dot* (New York: Random House, 1994).

2. Katherine Hignett, "Biblical City of Sodom Was Blasted to Smithereens by a Massive Asteroid Explosion," *Newsweek*, November 22, 2018, https://www.newsweek.com/biblical-city-sodom-was-blasted -smithereens-massive-asteroid-explosion-1227339 (accessed November 25, 2018).

3. D. Cox and J. Chestek, *Doomsday Asteroid: Can We Survive?* (Amherst, NY: Prometheus Books, 1996).

4. Wikipedia, s.v. "List of Impact Craters on the Earth," last modified December 26, 2018, 9:14, https://en.wikipedia.org/wiki/List_of_impact_craters_on_Earth.

5. Walter Alvarez, *T. Rex and the Crater of Doom* (Princeton, NJ: Princeton University Press, 1997).

CHAPTER 12. FOR OUR FREEDOM

1. Robert Zubrin, *Merchants of Despair* (New York: New Atlantis Books, 2012).

2. Friedrich von Bernhardi, *Germany and the Next War*, trans. Allen H. Powles (New York: Longmans, Green, 1914). German edition published in 1912.

3. Daniel Goldhagen, *Hitler's Willing Executioners* (New York: Vintage, 2007).

4. Lizzie Collingham, *The Taste of War: World War II and the Battle for Food* (New York: Penguin, 2012).

5. Zubrin, *Merchants of Despair*, chap. 6.

6. Timothy Snyder, "The Next Genocide," *New York Times*, September 12, 2015, https://www.nytimes.com/2015/09/13/opinion/sunday/the-next-genocide.html (accessed November 6, 2018); Timothy Snyder, *Black Earth: The Holocaust as History and Warning* (New York: Tim Duggan Books, 2015).

7. Michael Klare, *Resources Wars: The New Landscape of Global Conflict* (New York: Macmillan, 2001).

8. Robert Zubrin, *Energy Victory: Winning the War on Terror by Breaking Free of Oil* (Amherst, NY: Prometheus Books, 2007).

9. Robert Zubrin, "The Eurasianist Threat," *National Review*, March 3, 2014, https://www.nationalreview.com/2014/03/eurasianist-threat-robert-zubrin/ (accessed November 6, 2018).

10. Robert Zubrin, "America Stop Breathing," *National Review*, October 31, 2013, https://www.nationalreview.com/2013/10/america-stop-breathing-robert-zubrin/ (accessed November 6, 2018).

11. Robert G. Ingersoll, "Indianapolis Speech, 1876: Delivered to the Veteran Soldiers of the Rebellion," https://infidels.org/library/historical/robert_ingersoll/indianapolis_speech76.html (accessed November 6, 2018).

CHAPTER 13. FOR THE FUTURE

1. Quoted in "Konstantin Tsiolkovsky," URANOS, http://web.archive .org/web/20060421175318/http://www.uranos.eu.org/biogr/ciolke.html (accessed November 6, 2018, via Internet Archive WayBack Machine).

2. Jules Verne, *From the Earth to the Moon*, trans. Lowell Bair (New York: Bantam, 1993).

3. Vladimir I. Vernadsky, *Scientific Thought as a Planetary Phenomenon* (Moscow: Nongovernmental Ecological V. I. Vernadsky Foundation, 1997), http://vernadsky.name/wp-content/uploads/2013/02/ Scientific-thought-as-a-planetary-phenomenon-V.I2.pdf (accessed November 26, 2018).

CHAPTER 14. WHAT NEEDS TO BE DONE

1. Robert Zubrin, "A Purpose Driven Space Program?" *National Review*, March 2, 2018, https://www.nationalreview.com/2018/03/ nasa-lunar-space-station-unnecessary-space-flight-plans-lack-purpose/ (accessed November 6, 2018).

2. Robert Zubrin, "Cancel the Lunar Orbit Tollbooth," *National Review*, September 13, 2018, https://www.nationalreview.com/2018/09/nasa -lunar-orbiting-platform-gateway-should-be-canceled/ (accessed November 26, 2018).

INDEX

Page numbers in **bold** indicate figures, charts, tables, and plates.